OXFORD MATHEMATICAL MONOGRAPHS

OXFORD MATHEMATICAL MONOGRAPHS

A. Belleni-Moranti: *Applied semigroups and evolution equations*
J.W.P. Hirschfeld: *Projective geometries over finite fields*
N. Woodhouse: *Geometric quantization*
A.M. Arthurs: *Complementary variational principles* Second edition
P.L. Bhatnagar: *Nonlinear waves in one-dimensional dispersive systems*
N. Aronsajn, T.M. Creese, and L.J. Lipkin: *Polyharmonic functions*
J.A. Goldstein: *Semigroups of linear operators*
M. Rosenblum and J. Rovnyak: *Hardy classes and operator theory*
J.W.P. Hirschfeld: *Finite projective spaces of three dimensions*
K. Iwasawa: *Local class field theory*
A. Pressley and G. Segal: *Loop groups*
J.C. Lennox and S.E. Stonehewer: *Subnormal subgroups of groups*
D.E. Edmonds and W.D. Evans: *Spectral theory and differential operators*
Wang Jianhua: *The theory of games*
S. Omatu and J.H. Seinfeld: *Distributed parameter systems: theory and applications*
D. Holt and W. Plesken: *Perfect groups*
J. Hilgert, K.H. Hofmann, and J.D. Lawson: *Lie groups, convex cones, and semigroups*
S. Dineen: *The Schwarz lemma*
B. Dwork: *Generalized hypergeometric functions*
R.J. Baston and M.G. Eastwood: *The Penrose transform: its interaction with representation theory*
S.K. Donaldson and P.B. Kronheimer: *The geometry of four-manifolds*
T. Petrie and J. Randall: *Connections, definite forms, and four-manifolds*
R. Henstock: *The general theory of integration*
D.W. Robinson: *Elliptic operators and Lie groups*
A.G. Werschulz: *The computational complexity of differential and integral equations*
J.B. Griffiths: *Colliding plane waves in general relativity*
P.N. Hoffman and J.F. Humphreys: *Projective representations of the symmetric groups*
I. Györi and G. Ladas: *The oscillation theory of delay differential equations*
J. Heinonen, T. Kilpelainen, and O. Martio: *Non-linear potential theory*
B. Amberg, S. Franciosi, and F. de Giovanni: *Products of groups*
M.E. Gurtin: *Thermomechanics of evolving phase boundaries in the plane*
I. Ionescu and M. Sofonea: *Functional and numerical methods in viscoplasticity*
U. Grenander: *General pattern theory*
J. Faraut and A. Koranyi: *Jordan algebras and symmetric domains*
I.G. Macdonald: *Symmetric functions and Hall polynomials* 2nd edition
B.L.R. Shawyer and B.B. Watson: *Borel's methods of summability*

Borel's Methods of Summability

Theory and Applications

BRUCE SHAWYER and BRUCE WATSON

Department of Mathematics and Statistics
Memorial University of Newfoundland
St John's, Newfoundland, Canada

CLARENDON PRESS • OXFORD
1994

This book has been printed digitally in order to ensure its continuing availability

OXFORD
UNIVERSITY PRESS

Great Clarendon Street, Oxford OX2 6DP

Oxford University Press is a department of the University of Oxford.
It furthers the University's objective of excellence in research, scholarship,
and education by publishing worldwide in

Oxford New York

Auckland Bangkok Buenos Aires Cape Town Chennai
Dar es Salaam Delhi Hong Kong Istanbul Karachi Kolkata
Kuala Lumpur Madrid Melbourne Mexico City Mumbai Nairobi
São Paulo Shanghai Singapore Taipei Tokyo Toronto

with an associated company in Berlin

Published in the United States by Oxford University Press Inc., New York

Reprinted 2002

A catalogue record for this book is available from the British Library

Library of Congress Cataloging in Publication Data

(Data available)

ISBN 0 19 853585 6

to Jo and to Maureen,
for their loving support.

Preface

It has been ninety-two years since E. Borel published his *Leçons sur les séries divergentes* in which he first introduced his 'method' and just over a century since E. Cesaro initiated summability theory as a field of study. During that time the area of classical summability, as opposed to the functional analytic side, has matured to the point where relatively few new papers appear annually. Among the many different summability methods that have arisen over this time, Borel's have shown a justifiable longevity. They have proven to be particularly useful from the point of view of applications.

In light of the current state of the discipline we eagerly accepted the suggestion of Oxford University Press that publication of a survey book on Borel's methods would be timely. What we did not entirely realize when we started was how extensive the use of Borel's methods had become. Partly for this reason, and partly because of our reluctance to exclude things, the size of this book has doubled during the time of its creation.

Two of the reference works which have impressed us are the 'bible' of summability theory, *Divergent Series* by G. H. Hardy, and *Theorie der Limitierungsverfahren* by K. Zeller and W. Beekmann. No serious researcher could fail to appreciate the Chapter Notes provided by Hardy in the former volume and the extensive bibliography of the latter. Zeller and Beekman attempted to give a complete reference to the summability literature up to the publication date of their book. Here we have tried to do something similar for Borel summability at least as far as the 'Articles' and 'Applied Articles' sections are concerned. The section on 'Textbooks, Theses, and Lecture Notes' is incomplete but we hope we have not missed too many of the journal papers. In keeping with the spirit of the two volumes mentioned above we have attempted, in this book, to do three things. These are to place Borel's methods in the historical context in which they arose, to provide as complete coverage as possible of what we believe are the important results and to provide thorough references to the literature. We hope that the bibliography will play the dual role of providing additional details of results we have included and pointing the reader in the general direction of those we have missed.

Acknowledgement

The authors would like to acknowledge the contributions of their doctoral supervisor, Professor David Borwein, who introduced them to the subject of summability, and who has continued to encourage them throughout their careers.

The authors would like to thank Linda Bartholomew, Tracy Booth and Maitreyi Kothandaraman, summer research assistants at Memorial University for extensive bibliographic research work. They would also like to thank Wanda Heath for fast, cheerful, efficient, and accurate secretarial assistance.

Contents

0
Introduction

Here we collect the chapter headings in the rest of the book and give a brief summary of the contents of each chapter.

1 HISTORICAL OVERVIEW

A **very** brief survey of the subject of summability of series, ranging from the earliest attempts to sum convergent infinite series up to the first attempts to formalize the theory of summability of divergent series. For some of the material here we rely on a set of unpublished lecture notes of L. S. Bosanquet T[5].

2 SUMMABILITY METHODS IN GENERAL

Here we give the common types of summability methods, together with necessary and sufficient conditions for a summability method of one of these types to be *regular*; that is, for the method to assign the same sum to a convergent series. Some properties of specific methods used in the sequel are developed.

3 BOREL'S METHODS OF SUMMABILITY

This chapter begins with Borel's attempts to define two equivalent methods of summability. The two methods, now known as the B and the B' methods (named for Borel), are defined. Their properties are developed and some of their relationships with other known methods of summability are investigated.

4 RELATIONS WITH THE FAMILY OF CIRCLE METHODS

This chapter begins with a survey of the family of circle methods, as developed by Meyer–König. These methods are related to Borel's method, in that, under certain order conditions on the sequence $\{s_n\}$, they are essentially equivalent. We begin with Knopp's version of Euler's method, as first developed by Knopp in 1922. We then survey the Taylor methods, defined by Hardy and Littlewood in 1916. Included in the survey is Valiron's method, given by Hardy and Littlewood.

5 GENERALIZATIONS OF BOREL'S METHODS

All the known generalizations of Borel's methods are given here: generalizations based on Mittag-Leffler's functions as well as the strong and absolute summability methods corresponding to the originals.

6 ABELIAN THEOREMS.
Here we give results for which the prototype is Abel's theorem on the limit of a convergent power series at the end point of the interval of convergence. That is, theorems in which summability of one type necessarily implies summability of a second type.

7 and 8 TAUBERIAN THEOREMS I AND II
Tauberian theorems indicate the extra conditions required to proceed from an assumption of summability of one type to a conclusion of convergence or, more generally, to summability by a second type.

9 RELATIONSHIPS WITH OTHER METHODS
Borel's methods are not compatible with all summability methods. Here we examine how Borel's methods relate to other methods.

10 APPLICATIONS OF BOREL'S METHODS
There are many papers and books in the literature that make use of various properties of Borel's methods. We give a brief introduction to these.

11 REFERENCES
First, we give a listing of what we hope are most of the standard texts on classical summability theory. Some of these standard texts have whole sections on Borel's methods whereas others only mention them in passing. Next, we give a list of papers on Borel's methods, up to 1989, as well as other papers containing results required in some of the proofs presented here. Finally, we give a list of the papers in theoretical physics where use has been made of Borel's (or closely allied) methods. There are over 400 items in the bibliography.

1
Historical overview

Infinite series, whether used explicitly or implicitly, have fascinated mathematicians for centuries. In fact, convergence was familiar to Archimedes as early as 250 – 200 BC. He found the sum $s = 1 + \frac{1}{4} + \frac{1}{4^2} + \frac{1}{4^3} + \cdots$ by various methods of exhaustion. The simplest was by repeatedly quartering the unit square. Starting with a unit square (Fig. 1.1) we have:

$$\begin{aligned}1 &= \frac{3}{4} + \frac{1}{4} \\ &= \frac{3}{4} + \frac{1}{4}\left(\frac{3}{4} + \frac{1}{4}\right) \\ &= \frac{3}{4}\left(1 + \frac{1}{4}\right) + \frac{1}{4^2}\left(\frac{3}{4} + \frac{1}{4}\right) \\ &\vdots \\ &= \frac{3}{4}\left(1 + \frac{1}{4} + \frac{1}{4^2} + \cdots + \frac{1}{4^{n-1}}\right) + \frac{1}{4^n}.\end{aligned}$$

He then argued that the sum, s, could only satisfy $1 = \frac{3}{4}s$ or $s = \frac{4}{3}$.

This fact was used by Archimedes in his calculation of the area between a parabola and one of its chords. He proceeded roughly as follows. Construct the tangent to the curve parallel to the given chord AB. It touches the curve at C, say. Let Δ represent the area of triangle ABC. Next construct tangents parallel to AC and BC meeting the curve at D and E,

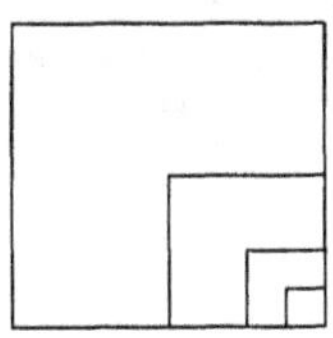

Fig. 1.1. Archimedes dissection of the unit square

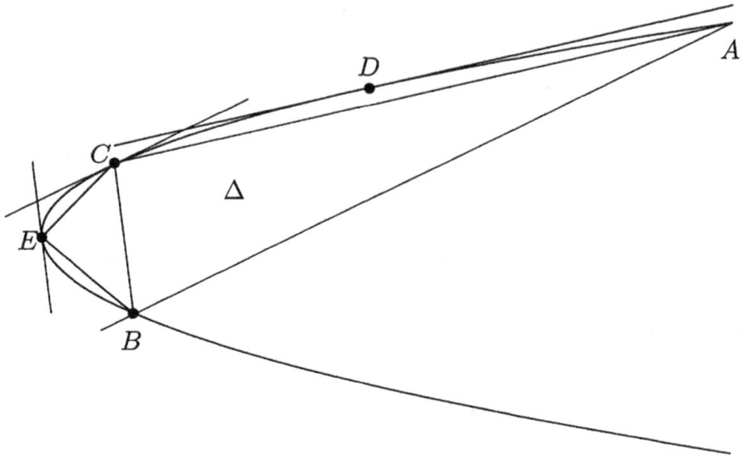

Fig. 1.2. Archimedes' argument for a parabola

respectively. It is an exercise to show that triangles BCE and ADC have total area $\frac{1}{4}\Delta$. Continuing the construction in this way we get, after n steps, that the total area of constructed triangles is

$$\Delta\left(1+\frac{1}{4}+\cdots+\frac{1}{4^{n-1}}\right).$$

Thus, Archimedes concluded that the total area was $\frac{4}{3}\Delta$. For a demonstration, see Simmons **T**[26], Section B.3.

Prior to the time of Cauchy and, in particular, to the publication of his book '*Cours d'Analyse Algébriques*' in 1821, there was no foundation for a general theory of convergence. However, the idea of convergence was familiar to Euler, Newton, and Leibnitz, even though Newton, for example, did not know a single test for convergence. They had an intuitive idea of the sum of a series as opposed to a formal definition of 'sum'.

But after Newton and Leibnitz, mathematicians were not so interested in the foundations of analysis since many problems could be tackled using the new methods of differential and integral calculus. Theorems using infinite series were obtained without regard to convergence since the results could often be verified by other means. It was similar to solving a differential equation by some operational method and verifying the solution directly afterward.

As an example of this, suppose we start with the series

(a) $\dfrac{1}{1-x} = 1 + x + x^2 + x^3 + \cdots$

obtained by long division, say. It was routine at that time to set $x = e^{i\theta}$ in (a) to get

(b) $1 + e^{i\theta} + e^{2i\theta} + e^{3i\theta} + \cdots = \left(1 - e^{i\theta}\right)^{-1} = \dfrac{1}{2} + \dfrac{1}{2} i \cot\left(\dfrac{\theta}{2}\right)$, for $0 < \theta < 2\pi$.

This gives the two equations

(c) $\dfrac{1}{2} + \cos(\theta) + \cos(2\theta) + \cos(3\theta) + \cdots = 0,$ and

(d) $\sin(\theta) + \sin(2\theta) + \sin(3\theta) + \cdots = \dfrac{1}{2}\cot\left(\dfrac{\theta}{2}\right).$

Translating to $[-\pi, \pi]$ gives

(e) $\dfrac{1}{2} - \cos(\theta) + \cos(2\theta) - \cos(3\theta) + \cdots = 0,$ and

(f) $\sin(\theta) - \sin(2\theta) + \sin(3\theta) - \cdots = \dfrac{1}{2}\tan\left(\dfrac{\theta}{2}\right).$

Integrating (e) term by term gives

$$\int_0^\theta \sum_{n=1}^{\infty} (-1)^{n+1} \cos n(\phi)\, d\phi = \int_0^\theta \frac{1}{2}\, d\phi, \quad \text{or}$$

(g) $\sin(\theta) - \dfrac{\sin(2\theta)}{2} + \dfrac{\sin(3\theta)}{3} - \cdots = \dfrac{\theta}{2}$ for $-\pi < \theta < \pi$,

a valid formula obtainable by other means.

Here, we have started with a valid formula, (a), and ended with a valid formula, (g), so there ought to be some theory to justify the transformations in between. Any such theory will necessarily involve a generalized definition of sum, for putting $\theta = 0$ in (e) gives

$$1 - 1 + 1 - 1 + \cdots = \frac{1}{2}.$$

Suspicion of these methods was best articulated by d'Alembert in the famous quote from his letter of 1768 in which he wrote that any calculations based on series which did not converge were very suspect. Of course, he was right! After limits were better understood and Cauchy, in his book, had articulated what we now refer to as the Cauchy Convergence Criterion, interest in the theory of infinite series revived.

The earliest interest in divergent series appears to have been initiated by Gauss. It was observed that divergent series did not behave in exactly the same way as convergent series. Attempts were made to assign a 'sum' to divergent series. For example, the series $\sum_{n=0}^{\infty}(-1)^n$, if treated like a convergent series with sum s, can lead to the following:

$$\begin{aligned} s &= 1-1+1-1+1-1+\cdots \\ &= (1-1)+(1-1)+(1-1)+\cdots \\ &= 0+0+0+\cdots = 0\,. \end{aligned}$$

But also

$$\begin{aligned} s &= 1-(1-1)+(1-1)+(1-1)+\cdots \\ &= 1-0+0+0+\cdots = 1\,, \end{aligned}$$

therefore $0 = 1$! This is, of course, a contradiction.

Even worse, we note that

$$\begin{aligned} s &= 1-1+1-1+1-1+\cdots \\ &= 1-(1-1)+(1-1)+(1-1)+\cdots \\ &= 1-s\,, \end{aligned}$$

so that $s = 1 - s$ or $s = \frac{1}{2}$.

All of this indicates a real need to deal carefully with divergent series.

It is generally accepted that the theory of divergent series began with an 1890 paper by Cesàro [64] on multiplication of series. In it he gave the first 'generalized definition' of the sum of a series. He considered the sequence of averages of partial sums of the given series. If that sequence had a limit he assigned the limit as the sum of the series, which he then referred to as a 'simply-indeterminate' series. This process is what we now call the $(C,1)$ method of summability.

Definition 1.1. *Let* $\sum_{n=1}^{\infty} a_n$ *be an infinite series of real or complex numbers with an associated sequence of partial sums* $\{s_n\}$. *Let* $\{t_n\}$ *represent the sequence of arithmetic means of* $\{s_n\}$ *so that:*

$$t_n = \frac{s_1 + s_2 + \cdots + s_n}{n}\,.$$

If $\lim_{n\to\infty} t_n = s$ *exists then we say that* $\sum_{n=1}^{\infty} a_n$ *is summable by the Cesàro method of order one with sum,* s*, and write* $\sum_{n=1}^{\infty} a_n = s(C,1)$.

Cesàro's definition has the advantage that for series which possess an 'expected sum' in the ordinary sense it is always the case that the series is $(C,1)$ summable and that 'Cesàro's sum' = 'Expected sum'. This fact, referred to as regularity for the method, is easily established as follows.

Let $\{s_n\}$ be a convergent sequence. We may assume without loss of generality that the sum is zero. Since $\lim_{n\to\infty} s_n = 0$, given $\epsilon > 0$, we can find $N = N(\epsilon) > 0$ such that $|s_n| < \epsilon$ whenever $n > N$. For $n \leq N$, we let $M = \sup\{|s_n| \ : \ n \leq N\}$, which is finite. Thus, for $n > M$,

$$
\begin{aligned}
|t_n| &= \left|\frac{s_1 + s_2 + \cdots + s_n}{n}\right| \\
&= \left|\frac{s_1 + \cdots + s_N + s_{N+1} + \cdots + s_n}{n}\right| \\
&\leq \frac{|s_1| + \cdots + |s_N| + |s_{N+1}| + \cdots + |s_n|}{n} \\
&\leq \frac{M + \cdots + M + \epsilon + \cdots + \epsilon}{n} \\
&= \frac{NM + (n-N)\epsilon}{n} \\
&\leq \frac{NM}{n} + \left(1 - \frac{N}{n}\right)\epsilon \\
&\leq \frac{NM}{n} + \epsilon \\
&\leq 2\epsilon \ \text{ for } \ n \text{ sufficiently large.}
\end{aligned}
$$

Hence, we deduce that $t_n \to 0$ as $n \to \infty$.

However, Cesaro's method will assign a sum to some series which are not ordinarily convergent. For $n \geq 1$, if $s_n = (-1)^{n+1}$, it is clear that $t_{2n} = 0$ and $t_{2n-1} = \frac{1}{n}$. Thus $t_n \to 0$ as $n \to \infty$.

Some of the impetus which led to the development of the theory of divergent series was provided by the following two results.

Cesàro's 1890 paper mentioned earlier dealt with multiplication of series. It was known that the Cauchy product of two convergent series need not converge. However, Mertens had shown that if one of the series converged absolutely then the product converged to the product of the sums.Cesàro established that the Cauchy product of two convergent series was 'simply indeterminate' to the expected sum; that is, the product of the sums.

There is Fejér's celebrated result on Fourier series. It was known that the Fourier series of a function, $f(x)$, need not converge to $f(x)$ even if f was continuous at x. Fejér showed, *inter alia*, that if f was continuous at x its Fourier series was $(C, 1)$ summable at x to $f(x)$.

The basis for the subject matter of this book, that is, Borel's exponential method, (B), was formulated in his first paper (December 1895) published in *Comptes Rendus* and entitled 'Sur la sommation des séries divergentes' [21]. A more detailed exposition appeared later in his book '*Leçons sur des Séries Divergentes*' which was published in 1901 **T**[2] with a second expanded edition in 1928 **T**[3]. Borel also defined what are now called power series methods of summability and extracted his exponential method as a special case. The regularity of the (B) method was given by Hardy in 1904 [104]. The basic definitions and results involving Borel's methods and generalizations are given in chapters 3 and 5 of this book.

2

Summability methods in general

2.1 Regularity

A summability method for a series $\sum_{n=0}^{\infty} a_n$ is simply any scheme for assigning to it a unique number which we then call the 'sum' of the series or the 'limit' of the associated sequence of partial sums $\{s_n\}$. In practice, working with a method can usually be reduced to consideration of one of the following transformations.

$$\begin{aligned}
&\textit{Sequence-to-sequence:} \quad & t_n &= \sum_{k=0}^{\infty} c_{nk}\, s_k. \\
&\textit{Sequence-to-function:} \quad & t(x) &= \sum_{k=0}^{\infty} c_k(x)\, s_k \quad \text{for} \quad 0 < x < \rho. \\
&\textit{Series-to-sequence:} \quad & y_n &= \sum_{k=0}^{\infty} d_{nk}\, a_k. \\
&\textit{Series-to-function:} \quad & y(x) &= \sum_{k=0}^{\infty} d_k(x)\, a_k \quad \text{for} \quad 0 < x < \rho.
\end{aligned}$$

Here, ρ may be finite or infinite.

For example, the $(C, 1)$ method of the previous chapter is a sequence-to-sequence method determined by a matrix transformation in which the doubly infinite matrix (c_{nk}) is given by:

$$c_{nk} = \begin{pmatrix} 1 & 0 & 0 & 0 & \cdots \\ \frac{1}{2} & \frac{1}{2} & 0 & 0 & \cdots \\ \frac{1}{3} & \frac{1}{3} & \frac{1}{3} & 0 & \cdots \\ \vdots & \vdots & \vdots & \vdots & \end{pmatrix}.$$

The matrix (c_{nk}) or the sequence of functions $\{c_k(x)\}$ is often called the *kernel* of the transformation.

A sequence-to-sequence or sequence-to-function method is said to be *regular* if $s_n \to s$ as $n \to \infty$ implies that the corresponding t_n or $t(x)$ tends to s as $n \to \infty$ or $x \to \rho^-$ respectively. Similarly, a series-to-sequence or series-to- function method is *regular* if $\sum_{n=0}^{\infty} a_n = s$ implies that the corresponding y_n or $y(x)$ tends to s as $n \to \infty$ or $x \to \rho^-$.

Summability theory is also used to assign values to 'improper' integrals. For summability of integrals, a fifth type of transformation arises.

Integral transformation: $t(x) = \int_0^{\nu} K(x, t) f(t)\, dt$ for $0 < x < \rho$

Here, we may regard $f(t)$ as $\int_0^t a(u)\, du$, where $0 < t < \nu$, and the integrand $a(u)$ is improper in some sense at ν.

Then, as with series, we assign the number L to the improper integral $\int_0^{\nu} a(u)\, du$ if $\lim_{x \to \rho^-} t(x) = L$. The integral transform determined by the *kernel* $K(x, t)$ is regular if $\lim_{t \to \nu^-} f(t) = L$ always implies that $\lim_{x \to \rho^-} t(x) = L$.

The main results of this chapter give conditions on the kernel under which a transformation will be regular. We give here only those results needed in the next chapters to deduce regularity of transforms related to Borel summability. For a thorough treatment of this topic see Hardy **T**[17].

The following result was obtained by Toeplitz [280] in 1911 for a sequence-to-sequence triangular transformation. It was extended, independently, by Kojima (1917) [155] and Schur (1920) [243].

Theorem 2.1. (the Toeplitz–Kojima–Schur theorem (TKS)). *Let* $\{c_k(x)\}$ *be a sequence of functions defined for* $x > 0$ *and* $k = 0, 1, 2, \cdots$. *Then* (A) *and* (B) *below hold if and only if* (I), (II), *and* (III) *below hold.*

(A) $\{s_k\}$ *convergent implies* $t(x) := \sum_{k=0}^{\infty} c_k(x)\, s_k$ *is convergent for each* $x > 0$.

(B) $\lim_{k \to \infty} s_k = s$ *exists implies* $\lim_{x \to \infty} t(x) = s$.

(I) $\sum_{k=0}^{\infty} c_k(x)$ *is absolutely convergent for each* $x > 0$ *and there exist* A, M *such that* $\sum_{k=0}^{\infty} |c_k(x)| < M$ *for all* $x > A \geq 0$.

(II) $c_k(x) \to 0$ *as* $x \to \infty$ *for each* $k = 0, 1, 2, \cdots$.

(III) $\sum_{k=0}^{\infty} c_k(x) \to 1$ *as* $x \to \infty$.

Specifying $c_k(x) = c_{nk}$ when $n \le x < n+1$ gives the sequence-to-sequence version as found in, say, Hardy **T**[17].

First some preliminary lemmas.

Lemma 2.2. *The following are equivalent.*

(i) $\sum_{k=0}^{\infty} c_k(x)$ *is absolutely convergent for each* $x > 0$.

(ii) $\sum_{k=0}^{\infty} c_k(x)\, s_k$ *converges for each* $x > 0$ *whenever* $s_k \to s$.

(iii) $\sum_{k=0}^{\infty} c_k(x)\, s_k$ *converges for each* $x > 0$ *whenever* $s_k \to 0$.

Proof. Clearly (i) $\to$ (ii) $\to$ (iii). So assume (iii) and suppose, by way of contradiction, that $\sum_{k=0}^{\infty} |c_k(x)| = \infty$.

Then, there exists a subsequence $\{n_i\}$ of the nonnegative integers such that

$$\sum_{k=n_i}^{n_{i+1}-1} |c_k(x)| > i \quad \text{for} \quad i = 1, 2, 3, \cdots.$$

Define a sequence $\{s_k\}$ as follows.

$$s_k = \begin{cases} \dfrac{1}{i}\dfrac{|c_k(x)|}{c_k(x)} & \text{if} \quad c_k(x) \ne 0,\ n_i \le k < n_{i+1} \\ 0 & \text{if} \quad c_k(x) = 0 \quad \text{or} \quad 0 \le k < n_1 \end{cases}$$

Then $s_k \to 0$ but

$$\sum_{k=n_i}^{n_{i+1}-1} c_k(x) s_k = \sum_{k=n_i}^{n_{i+1}-1} \frac{1}{i} |c_k(x)|$$

$$= \frac{1}{i} \sum_{k=n_i}^{n_{i+1}-1} |c_k(x)| > 1.$$

Hence $\sum_{0}^{\infty} c_k(x) s_k$ must diverge, which is a contradiction.

Lemma 2.3. *$c_n(x) \to 0$ as $x \to \infty$ if and only if $\sum_0^\infty c_k(x)s_k \to 0$ as $x \to \infty$ whenever $s_k = \begin{cases} 1 & k = n \\ 0 & k \neq n \end{cases}$ for each $n = 0, 1, 2, \cdots$.*

Proof. Here, $t(x)$ is simply $c_n(x)$ so the result is trivial.

Lemma 2.4. *$\sum_{k=0}^\infty c_k(x) \to 1$ as $x \to \infty$ if and only if $\sum_{k=0}^\infty c_k(x)\, s_k \to 1$ as $x \to \infty$ where $s_k = 1$ for all $k \geq 0$.*

Proof. $t(x) = \sum_{k=0}^\infty c_k(x)$ so again the result is trivial.

Lemma 2.5. (see Schur [243]). (I) *and* (II) *hold if and only if $s_k \to 0$ implies $\sum_{k=0}^\infty c_k(x)s_k$ exists and tends to zero as $x \to \infty$.*

Proof. Suppose (I) and (II) hold. Then $t(x)$ exists by lemma 2.2. Moreover, for any fixed n,

$$\begin{aligned}\limsup_{x\to\infty} |t(x)| &= \limsup \left|\sum_{k=0}^\infty c_k(x)s_k\right| \\ &\leq \limsup \sum_{k=n}^\infty |c_k(x)|\,|s_k| \\ &\leq M \sup_{k\geq n} |s_k|.\end{aligned}$$

Hence $t(x) \to 0$ whenever $s_k \to 0$.

Consider the reverse implication. By lemma 2.3 we get that (II) holds, and $\sum_{k=0}^\infty c_k(x)$ is absolutely convergent for $x > 0$ by lemma 2.2. It remains to show that $\sum_{k=0}^\infty |c_k(x)|$ is uniformly bounded from some point on. Suppose, by way of contradiction, that this is not the case. That is, assume for any pair M, A of positive numbers, that there exists $x > A$ such that $\sum_{k=0}^\infty |c_k(x)| > M$. We proceed to construct a null sequence $\{s_k\}$ for which $t(x)$ fails to converge to 0.

Let $\{M_i\}$ be a fixed sequence of integers satisfying $2 < M_1 < M_2 < \cdots$. Inductively, construct sequences $\{x_i\}, \{A_i\}$, and a subsequence $\{n_i\}$ of the natural numbers as follows. Let $x_0 > 0$ and $n_1 > 0$ be given. Set $s_k = 0$ if $0 \le k < n_1$. Since $\sum_{k=0}^{\infty} |c_k(x)| < \infty$ for all $x > 0$ and (II) holds, there is an $A_1 > x_0$ such that $\sum_{k=0}^{n_1-1} |c_k(x)| < 1$ for all $x > A_1$. By assumption, there is an $x_1 > A_1 + 1$ such that $\sum_{k=0}^{\infty} |c_k(x)| > M_1$. But $\sum_{k=0}^{\infty} |c_k(x_i)| < \infty$. Hence, there is an $n_2 > n_1$ such that $\sum_{k=n_1-1}^{\infty} |c_k(x_1)| < 1$. Note that $\sum_{k=n_1}^{n_2-1} |c_k(x_1)| = \left(\sum_{k=0}^{\infty} - \sum_{k=0}^{n_1-1} - \sum_{k=n_2}^{\infty}\right) |c_k(x_1)| > M_1 - 1 - 1 = M_1 - 2$. For $n_1 \le k < n_2$, set $s_k = \dfrac{1}{2}\dfrac{|c_k(x_1)|}{c_k(x_1)}$ if $c_k(x_1) \ne 0$ and $s_k = 0$ if $c_k(x_1) = 0$.

We continue in this fashion. There exists an $A_i > x_{i-1}$ such that $\sum_{k=0}^{n_i-1} |c_k(x)| < 1$ for all $x > A_i$. Then there is an $x_i > A_i + 1$ with $\sum_{k=0}^{\infty} |c_k(x_i)|$ such that $\sum_{k=n_{i+1}}^{\infty} |c_k(x_i)| < 1$. Moreover $\sum_{k=n_i}^{n_{i+1}-1} |c_k(x_i)| > M_i - 1 - 1 = M_i - 2$.

$$\text{Set } s_k = \begin{cases} \dfrac{1}{i+1}\dfrac{|c_k(x_i)|}{c_k(x_i)} & \text{if} \quad c_k(x_i) \ne 0,\ n_i \le k < n_{i+1} \\ 0 & \text{if} \quad c_k(x_i) = 0. \end{cases}$$

Then $|s_k| \le \frac{1}{i+1}$ for $n \ge n_i$ so $|s_k| \le 1$ for all $k \ge 0$ and $s_k \to 0$ as $k \to \infty$.

For any fixed i we have

$$\begin{aligned} |t(x_i)| &= \left|\sum_{k=0}^{\infty} c_k(x_i)s_k\right| \\ &\ge \left|\sum_{k=n_i}^{n_{i+1}-1} c_k(x_i)s_k\right| - \sum_{k=0}^{n_i-1} |c_k(x_i)s_k| - \sum_{k=n_{i+1}}^{\infty} |c_k(x_i)s_k| \\ &\ge \sum_{k=n_i}^{n_{i+1}-1} \frac{|c_k(x_i)|}{i+1} - 1 - 1 \\ &> \frac{M_i - 2}{i+1} - 2 \end{aligned}$$

Set $M_i = 3i + 5$. Then $|t(x_i)| > 1$ for any i. Hence we cannot have $t(x) \to 0$. This is a contradiction, which completes the proof.

Proof of Theorem 2.1 (TKS). Suppose (I), (II), and (III) hold. Then (A) holds by lemma 2.2. We may write

$$\begin{aligned} t(x) - s &= \sum_{k=0}^{\infty} c_k(x)s_k - s \\ &= \sum_{k=0}^{\infty} c_k(x)(s_k - s) + s\sum_{k=0}^{\infty} c_k(x) - s. \end{aligned}$$

Therefore $t(x) - s = \sum_{k=0}^{\infty} c_k(x)(s_k - s) + s\left(\sum_{k=0}^{\infty} c_k(x) - 1\right)$. The first term tends to zero as $x \to \infty$ by lemma 2.5 and (III) implies the second also converges to zero.

Suppose that (A) and (B) hold.

By lemma 2.4 and (B), (III) holds. Suppose $s_k \to 0$. Then (A) implies $t(x)$ exists and (B) forces $t(x) \to 0$.

Hence, by lemma 2.5, (I) and (II) hold. This completes the proof.

The above result is used to establish regularity of sequence-to-sequence and sequence-to-function methods. We will also need to establish regularity of integral transforms. The next theorem is the integral analogue of the TKS theorem. We give only the sufficient form of the result since that is all we require and the necessary and sufficient conditions for integrals are less tidy than those for sequences (see, for example, Dunford and Schwarz **T**[13], exercise 79, pp. 351–352).

We assume that $K(x,t)$ and $f(t)$ are Lebesgue measurable on $(0,\infty)$ as functions of t and that $f(t)$ is bounded (hence integrable) on every finite interval of the form $(0,X)$. In practice, f will be continuous on $(0,\infty)$. We regard $t(x)$ as a Cauchy–Lebesgue integral. That is, $t(x) := \lim_{T\to\infty} \int_0^T K(x,t)f(t)\,dt$ if the limit exists.

Theorem 2.6. *Suppose that*

(I) $\displaystyle\int_0^{\infty} |K(x,t)|\,dt < M$ *for all* $x > 0,$

(II) $\displaystyle\int_0^{Y} |K(x,t)|\,dt \to 0$ *as* $x \to \infty$ *for each fixed* $Y > 0,$ *and*

(III) $\displaystyle\int_0^{\infty} |K(x,t)|\,dt \to 1$ *as* $x \to \infty.$

Then $t(x)$ exists and tends to s as $x \to \infty$ whenever $f(t) \to s$ as $t \to \infty$.

Proof. Suppose that $f(t) \to 0$ as $t \to \infty$. Then there exists an M_0 such that $|f(t)| < M_0$ for all $t \in (0, \infty)$. Moreover, $t(x)$ exists as a Lebesgue integral over $(0, \infty)$ by virtue of (I).

Assign $\varepsilon > 0$ and write

$$t(x) = \left(\int_0^Y + \int_Y^\infty \right) K(x,t) f(t)\, dt = U(x) + V(x).$$

Since $f(t) \to 0$ as $t \to \infty$, there exists a Y_0 such that $|f(t)| < \frac{\varepsilon}{2} M$ for all $t > Y_0$.

Hence, $|V(x)| \le \dfrac{\varepsilon}{2M} \displaystyle\int_{Y_0}^\infty |K(x,t)|\, dt < \dfrac{\varepsilon}{2}$. By (II), there is an X such that $\displaystyle\int_0^{Y_0} |K(x,t)|\, dt < \dfrac{\varepsilon}{2M_0}$ for all $x > X$. Therefore, if $x > X$, we get $|U(x)| + |V(x)| < \varepsilon$ and hence $t(x) \to 0$.

Now suppose $f(t) \to s$ as $t \to \infty$ and write $F(t) = f(t) - s$. Then

$$t^*(x) := \int_0^\infty K(x,t) F(x)\, dt \to 0 \quad \text{as} \quad x \to \infty.$$

But,

$$\begin{aligned} t(x) &= \int_0^\infty K(x,t)\{F(x) + s\}\, dt \\ &= t^*(x) + s \int_0^\infty K(x,t)\, dt \\ &\to s \qquad \text{as} \qquad x \to \infty, \quad \text{by (III)}. \end{aligned}$$

This completes the proof.

The two main results of this section assume that the continuous parameters tend to infinity. This need not always be the case. We end this section by stating straightforward extensions of Theorems 2.1 and 2.6. We omit the proofs. They follow from suitable transformations of variables.

Theorem 2.7. *Let $\{c_k(x)\}$ be a sequence of functions defined for $0 < x < \rho$ and $k = 0, 1, 2, \cdots$. Then* (A) *and* (B) *below hold if and only if* (I), (II), *and* (III) *below hold.*

(A) *$\{s_k\}$ convergent implies $t(x) := \displaystyle\sum_{k=0}^\infty c_k(x)\, s_k$ is convergent for each $x \in (0, \rho)$.*

(B) $\lim_{k\to\infty} s_k = s$ *exists implies* $\lim_{x\to\rho^-} t(x) = s$.

(I) $\sum_{k=0}^{\infty} c_k(x)$ *is absolutely convergent for each* $x \in (0,\rho)$ *and there exist* A, M *such that* $0 < A < \rho$ *and* $\sum_{k=0}^{\infty} |c_k(x)| < M$ *for all* $x \in (A,\rho)$.

(II) $c_k(x) \to 0$ *as* $x \to \rho^-$ *for each* $k = 0, 1, 2, \cdots$.

(III) $\sum_{k=0}^{\infty} c_k(x) \to 1$ *as* $x \to \rho^-$.

Theorem 2.8. *Suppose that*

(I) $\int_0^{\nu} |K(x,t)|\, dt < M$ *for all* $x \in (0,\rho)$,

(II) $\int_0^{Y} |K(x,t)|\, dt \to 0$ *as* $x \to \rho^-$ *for each fixed* $Y \in (0,\nu)$, *and*

(III) $\int_0^{\nu} |K(x,t)|\, dt \to 1$ *as* $x \to \rho^-$.

Then $t(x)$ *exists and tends to* s *as* $x \to \rho^-$ *whenever* $f(t) \to s$ *as* $t \to \nu^-$.

In the theorem above we assume that the required measurability conditions hold on the appropriate intervals.

2.2 Generalized Cesàro summability

In this and subsequent sections in this chapter we define and set out the basic properties of some specific methods of summability. In later chapters, these methods will be compared to Borel methods.

This section deals with generalized Cesàro summability. The Cesàro 'averaging' method $(C,1)$ was defined in chapter 1. There are many ways to extend this method. First, we follow the treatment in Hardy **T**[17], for example, and develop a scale of methods (C,α) for $\alpha \geq 0$.

The process starts by inductively creating a sequence $\{A_n^\alpha\}$ as follows. Let $A_n^{-1} = a_n$, $A_n^0 = s_n = \sum_{k=0}^{n} a_k$ and, in general, $A_n^\alpha = \sum_{k=0}^{n} A_k^{\alpha-1}$ for $\alpha = 0, 1, 2, \cdots$. Additionally, let E_n^α represent A_n^α in the special case where $a_0 = 1$ and $a_n = 0$ for $n = 1, 2, \cdots$.

Definition 2.9. *For $\alpha = 0, 1, 2, \cdots$, we say that $\sum_{n=0}^{\infty} a_n$ is summable to s by the method (C, α) and write $\sum_{n=0}^{\infty} a_n = s\ (C, \alpha)$, or $s_n \to s\ (C, \alpha)$, if* $\lim_{n\to\infty} \frac{A_n^\alpha}{E_n^\alpha} = s$.

Note that $(C, 0)$ reduces to ordinary convergence and that $(C, 1)$ is identical to the definition of $(C, 1)$ already given in chapter 1.

Next, for $|x| < 1$, consider the following formal power series identities:

$$\begin{aligned}\frac{1}{1-x}\sum_{k=0}^{\infty} A_k^{\alpha-1}x^k &= \sum_{k=0}^{\infty} x^k \sum_{k=0}^{\infty} A_k^{\alpha-1}x^k \\ &= \sum_{k=0}^{\infty} A_k^{\alpha}x^k\end{aligned}$$

by the Cauchy product of series.

Thus, inductively,

$$\begin{aligned}\sum_{k=0}^{\infty} A_k^\alpha x^k &= \frac{1}{(1-x)}\sum_{k=0}^{\infty} A_k^{\alpha-1}x^k \\ &= \frac{1}{(1-x)^2}\sum_{k=0}^{\infty} A_k^{\alpha-2}x^k \\ &\ \ \vdots \\ &= \frac{1}{(1-x)^\alpha}\sum_{k=0}^{\infty} s_k x^k \\ &= \frac{1}{(1-x)^{\alpha+1}}\sum_{k=0}^{\infty} a_k x^k.\end{aligned}$$

This suggests that, for any $\alpha > -1$, we may simply define A_n^α and E_n^α to be the quantities satisfying

$$\sum_{k=0}^{\infty} E_k^\alpha x^k = \frac{1}{(1-x)^{\alpha+1}}, \quad \text{and} \tag{2.1}$$

$$\sum_{k=0}^{\infty} A_k^\alpha x^k = \frac{1}{(1-x)^\alpha}\sum_{k=0}^{\infty} s_k x^k = \frac{1}{(1-x)^{\alpha+1}}\sum_{k=0}^{\infty} a_k x^k. \tag{2.2}$$

Definition 2.10. *For $\alpha \geq 0$, a sequence $\{s_n\}$ or series $\sum_{n=0}^{\infty} a_n$ is summable (C, α) to s if $\lim_{n\to\infty} \frac{A_n^\alpha}{E_n^\alpha} = s$. We then write either $\sum_{n=0}^{\infty} a_n = s\ (C, \alpha)$, or $s_n \to s\ (C, \alpha)$.*

Some useful identities are summarized in the next lemma.

Lemma 2.11. *The following identities hold for $\alpha > -1$, $\beta > -1$, and $\alpha + \beta > -1$.*

(1) $E_n^\alpha = \binom{n+\alpha}{n} = \binom{n+\alpha}{\alpha} = \dfrac{(\alpha+1)(\alpha+2)\dots(\alpha+n)}{n!}$.

(2) $A_n^\alpha = \sum_{k=0}^{n} \binom{n-k+\alpha}{\alpha} a_k = \sum_{k=0}^{n} \binom{k+\alpha}{k} a_{n-k}$.

(3) $\binom{n+\alpha+\beta+1}{n} = \sum_{k=0}^{n} \binom{n-k+\alpha}{\alpha}\binom{k+\beta}{\beta}$.

(4) $A_n^{\alpha+\beta+1} = \sum_{k=0}^{n} A_k^\alpha E_{n-k}^\beta = \sum_{k=0}^{n} E_k^\beta A_{n-k}^\alpha$.

Proof. All follow from (2.1) and (2.2). Note that for $|x| < 1$, the series in (2.1) is convergent.

Next, we re-phrase the definition of (C, α) summability and extend it to the two-parameter version as introduced by Borwein [38].

For $\beta > -1$ and $\alpha + \beta > -1$, write

$$s_n^\alpha = \frac{1}{E_n^\alpha} \sum_{k=0}^{n} E_{n-k}^{\alpha-1} s_k, \quad \text{and}$$

$$s_n^{\alpha,\beta} = \frac{1}{E_n^{\alpha+\beta}} \sum_{k=0}^{n} E_{n-k}^{\alpha-1} E_k^\beta s_k.$$

Note that s_n^α is precisely A_n^α / E_n^α by lemma 2.11.

Definition 2.12. *Suppose $\beta > -1$ and $\alpha + \beta > -1$. Then $\sum_{n=0}^{\infty} a_n$ is said to be summable to s by the method (C, α, β) if $s_n^{\alpha,\beta} \to s$ as $n \to \infty$.*

Note that $(C, 0, 0)$ is ordinary convergence and that $(C, \alpha, 0) = (C, \alpha)$ for $\alpha > 0$. In particular, $(C, 1)$ is as defined earlier.

Theorem 2.13. *The method (C, α, β) is a regular summability method for $\alpha > 0$.*

Proof. Write $c_{nk} = \dfrac{E_{n-k}^{\alpha-1} E_k^{\beta}}{E_n^{\alpha+\beta}}$ for $0 \le k \le n$ and $c_{nk} = 0$ for $k > n$. Then, by lemma 2.11,

$$\begin{aligned}\sum_{k=0}^{n} |c_{nk}| &= \sum_{k=0}^{n} c_{nk} \\ &= \frac{1}{E_n^{\alpha+\beta}} \sum_{k=0}^{n} E_{n-k}^{\alpha-1} E_k^{\beta} \\ &= 1.\end{aligned}$$

Moreover, since $\dbinom{n+\alpha}{n} \sim \dfrac{n^{\alpha}}{\Gamma(\alpha+1)}$ we get that

$$c_{nk} \sim \frac{\Gamma(\alpha+\beta+1)}{\Gamma(\alpha)\Gamma(\beta+1)} \frac{1}{n} \to 0 \quad \text{as} \quad n \to \infty.$$

Hence, by Theorem 2.1, (C, α, β) is regular in the given range for α. This completes the proof.

For two summability methods P and Q we use the following standard definitions. If $\sum_{n=0}^{\infty} a_n = s(P)$ whenever $\sum_{n=0}^{\infty} a_n = s(Q)$ then P is said to *include* Q and we write $P \supseteq Q$. If $P \supseteq Q$ and there is a series $\sum_{n=0}^{\infty} a_n$ summable (to s) by P but not by Q then we write $P \supset Q$. If $P \supseteq Q$ and $Q \supseteq P$ then P and Q are said to be equivalent and we write $P \simeq Q$.

We complete this brief study of Cesàro's methods by giving an inclusion result for the (C, α) scale, and proving some lemmas used here and in subsequent chapters.

Theorem 2.14. *For $\alpha' > \alpha > -1$, $(C, \alpha') \supseteq (C, \alpha)$.*

Proof. Suppose $\sum_{n=0}^{\infty} a_n = s\,(C, \alpha)$. Then,

$$s_n^{\alpha'} = \frac{1}{E_n^{\alpha'}} \sum_{k=0}^{n} E_{n-k}^{\alpha'-\alpha-1} A_k^{\alpha}$$

$$= \frac{1}{E_n^{\alpha'}} \sum_{k=0}^{n} E_{n-k}^{\alpha'-\alpha-1} E_k^{\alpha} \left(\frac{A_k^{\alpha}}{E_k^{\alpha}} \right)$$

$$=: \sum_{k=0}^{\infty} c_{nk} s_k^{\alpha}.$$

But (c_{nk}) is the kernel of the $(C, \alpha' - \alpha, \alpha)$ transform since

$$\sum_{k=0}^{\infty} c_{nk} = \frac{1}{E_n^{\alpha'}} \sum_{k=0}^{n} E_{n-k}^{\alpha'-\alpha-1} E_k^{\alpha}$$

$$= \frac{1}{E_n^{\alpha'}} \sum_{k=0}^{n} E_k^{\alpha'-\alpha-1} E_{n-k}^{\alpha}.$$

The result follows since $(C, \alpha' - \alpha, \alpha)$ is regular by Theorem 2.13.

The next result may appear out of place here. However, it does involve the notation, ideas, and some of the results of this section. Moreover, it will be used later in certain theorems connecting Borel and Cesàro summability. It appears as theorem 144 in Hardy **T**[17]. First, recall one fact found in, say, Stromberg **T**[27] about the gamma function, $\Gamma(x)$. For $0 < \beta \le 1$ and $n = 1, 2, \cdots$ we always have

$$(n-1)^{\beta} \le \frac{\Gamma(n+\beta)}{(n-1)!} \le n^{\beta}. \tag{2.3}$$

Lemma 2.15. *Suppose that* $\alpha \ge -1$, $0 < \beta \le 1$, $\rho > -1$, $0 < H < 1$ *and* $A_n^{\alpha} = \mathbf{o}(n^{\rho})$. *Then*

(1) $A_n^{\alpha+\beta} = \mathbf{o}(n^{\rho+\beta})$, *and*

(2) $A_m^{\alpha+\beta} - A_n^{\alpha+\beta} = \mathbf{o}(|m-n|^{\beta} n^{\rho})$ *uniformly for* $0 < (1-H)n \le m \le (1+H)n$.

Proof. For part (1), it suffices to show that $A_n^{\alpha+\beta}/E_n^{\rho+\beta} \to 0$ as $n \to \infty$. But, as in theorem 2.14, we can write

$$\frac{A_n^{\alpha+\beta}}{E_n^{\rho+\beta}} = \frac{1}{E_n^{\rho+\beta}} \sum_{k=0}^{n} E_{n-k}^{\beta-1} E_k^{\rho} \left(\frac{A_k^{\alpha}}{E_k^{\rho}} \right).$$

The proof then follows since (C, β, ρ) is regular.

For part (2), write $\dfrac{\Gamma(n-k+\beta)}{\Gamma(\beta)\Gamma(n-k+1)}$ for $E_{n-k}^{\beta-1}$ and suppose that $m > n$. Then,

$$\begin{aligned}
&\Gamma(\beta)(A_m^{\alpha+\beta} - A_n^{\alpha+\beta}) \\
&= \sum_{k=0}^{m} \frac{\Gamma(m-k+\beta)}{\Gamma(m-k+1)} A_k^\alpha - \sum_{k=0}^{n} \frac{\Gamma(n-k+\beta)}{\Gamma(n-k+1)} A_k^\alpha \\
&= \sum_{k=n+1}^{m} \frac{\Gamma(m-k+\beta)}{\Gamma(m-k+1)} A_k^\alpha + \sum_{k=0}^{n} \left\{ \frac{\Gamma(m-k+\beta)}{\Gamma(m-k+1)} - \frac{\Gamma(n-k+\beta)}{\Gamma(n-k+1)} \right\} A_k^\alpha .
\end{aligned}$$

Let S_1 and S_2, respectively, denote the two sums in the line above.

Since $A_k^\alpha = \mathbf{o}(k^\rho)$ and $(1-H)n \leq k \leq (1+H)n$, we get that $A_k^\alpha = \mathbf{o}(n^\rho)$. Hence, by (2.3),

$$\begin{aligned}
\frac{|S_1|}{(m-n)^\beta n^\rho} &\leq \frac{1}{(m-n)^\beta} \sum_{k=n+1}^{m-1} (m-k)^{\beta-1} \mathbf{o}(1) + \frac{\Gamma(\beta)}{\Gamma(1)} \frac{A_m^\alpha}{n^\rho} \\
&= \mathbf{o}(1).
\end{aligned}$$

For S_2 assume, as we may, that $\beta < 1$ or else $S_2 = 0$. Then, using (2.3) again, we get

$$\begin{aligned}
|S_2| &= \left(\sum_{k=0}^{[\frac{n}{2}]} + \sum_{k=[\frac{n}{2}]+1}^{n} \right) \left\{ \frac{\Gamma(n-k+\beta)}{\Gamma(n-k+1)} - \frac{\Gamma(m-k+\beta)}{\Gamma(m-k+1)} \right\} \mathbf{o}(k^\rho) \\
&= S_3 + S_4, \quad \text{say.}
\end{aligned}$$

For S_4, $A_k^\alpha = \mathbf{o}(n^\rho)$ for k in the given range. Summing from 0 to n gives

$$\begin{aligned}
S_4 &= \mathbf{o}\left(n^\rho \sum_{k=0}^{n} \left\{ \frac{\Gamma(n-k+\beta)}{\Gamma(n-k+1)} - \frac{\Gamma(m-k+\beta)}{\Gamma(m-k+1)} \right\} \right) \\
&= \mathbf{o}\left(n^\rho \left(\sum_{k=0}^{n} - \sum_{k=m-n}^{m} \right) \frac{\Gamma(k+\beta)}{\Gamma(k+1)} \right).
\end{aligned}$$

But $0 < m-n \leq (1+H)n - n = Hn < n$. Therefore

$$S_4 = \mathbf{o}\left(n^\rho \left(\sum_{k=0}^{m-n-1} - \sum_{k=n+1}^{m} \right) \frac{\Gamma(k+\beta)}{\Gamma(k+1)} \right)$$

$$
\begin{aligned}
&= \mathbf{o}\left(n^\rho \sum_{k=0}^{m-n} \frac{\Gamma(k+\beta)}{\Gamma(k+1)}\right) \\
&= \mathbf{o}\left(n^\rho \sum_{k=0}^{m-n} \frac{1}{k+\beta}(k+1)^\beta\right) \\
&= \mathbf{o}\left(n^\rho \sum_{k=0}^{m-n} (k+1)^{\beta-1}\right) \\
&= \mathbf{o}\left(n^\rho (m-n)^\beta\right).
\end{aligned}
$$

Finally, for S_3 where $0 \le k \le \left[\frac{n}{2}\right]$, we have

$$
\frac{\Gamma(n-k+\beta)}{(n-k)!} - (n-k)^{\beta-1} = \mathbf{O}\left(n^{\beta-2}\right).
$$

Hence,

$$
\begin{aligned}
&\frac{\Gamma(n-k+\beta)}{\Gamma(n-k+1)} - \frac{\Gamma(m-k+\beta)}{\Gamma(m-k+1)} \\
&= (n-k)^{\beta-1} - (m-k)^{\beta-1} + \mathbf{O}\left(n^{\beta-2}\right) \\
&= \mathbf{O}\left((m-n)n^{\beta-2}\right) + \mathbf{O}\left(n^{\beta-2}\right) \\
&= \mathbf{O}\left((m-n)n^{\beta-2}\right).
\end{aligned}
$$

Therefore,

$$
\begin{aligned}
S_3 &= \mathbf{o}\left((m-n)n^{\beta-2} \sum_{k=0}^{\left[\frac{n}{2}\right]} k^\rho\right) \\
&= \mathbf{o}\left((m-n)n^{\beta+\rho-1}\right) \\
&= \mathbf{o}\left(\frac{(m-n)^{1-\beta}(m-n)^\beta n^\rho}{n^{1-\beta}}\right) \\
&= \mathbf{o}\left((m-n)^\beta n^\rho\right).
\end{aligned}
$$

This completes the proof.

2.3 Sequence-to-function methods based on power series

Suppose that $\{p_n\}$ is a sequence of nonnegative numbers with the property that $\sum_{k=n}^{\infty} p_k > 0$ for $n = 0, 1, 2, \cdots$. Denote the radius of convergence of $p(x) := \sum_{n=0}^{\infty} p_n x^n$ by ρ_p or, if there is no ambiguity about the underlying sequence, simply by ρ. For a sequence $\{s_n\}$ (or series $\sum a_n$), define $p_s(x)$ by $p_s(x) = \frac{1}{p(x)} \sum_{n=0}^{\infty} p_n s_n x^n$.

Definition 2.16. *Suppose that $\rho > 0$ and $p_s(x)$ exists for x in $(0, \rho)$. If $\lim_{x \to \rho^-} p_s(x) = s$ then we say that $\{s_n\}$ is summable by the power series method, (P), based on $\{p_n\}$ and write $s_n \to s$ (P).*

Theorem 2.17. (see Borwein [35]). *Suppose that (P) is a power series method based on $\{p_n\}$.*

(1) *If $\rho = \infty$ then (P) is regular.*

(2) *If $0 < \rho < \infty$ then (P) is regular if and only if $\sum_{n=0}^{\infty} p_n \rho^n = \infty$.*

Proof. We appeal to theorem 2.1 (TKS), and so set $c_k(x) = p_k x^k / p(x)$. Then we always have $\sum_{k=0}^{\infty} c_k(x) = 1$.

For (1) fix k and select, as we may, $m > k$ such that $p_m > 0$. Then

$$0 \leq c_k(x) \leq \frac{p_k x^k}{p_m x^m} \to 0 \quad \text{as } x \to \infty.$$

For (2) first observe that $\lim_{x \to \rho^-} p(x) = \sum_{n=0}^{\infty} p_n \rho^n$ since, for any $m \geq 0$,

$$\sum_{n=0}^{\infty} p_n \rho^n \geq \lim_{x \to \rho^-} p(x) \geq \sum_{k=0}^{m} p_k \rho^k$$

. It now follows that, for any fixed k, $\lim_{x \to \rho^-} c_k(x) = \lim_{x \to \rho^-} \frac{p_k x^k}{p(x)} = 0$ if and only if $\lim_{x \to \rho^-} p(x) = \sum_{k=0}^{\infty} p_k \rho^k = \infty$. This completes the proof.

The most important sequence-to-function method (well ... given the title of this book!) is the Borel summability method (B). It is a power series method based on the sequence $p_n = \frac{1}{n!}$. Using the observation that $p(x) = e^x$ and theorem 2.17 it is clear that the method (B) is regular. We formalize these observations as follows.

Definition 2.18. *A sequence $\{s_n\}$ (or its corresponding series $\sum_{n=0}^{\infty} a_n$) is said to be Borel summable to s if $e^{-x} \sum_{n=0}^{\infty} \frac{s_n}{n!} x^n$ exists for $x > 0$ and tends to s as x tends to ∞. In this case we write $s_n \to s$ (B).*

Theorem 2.19. *The method (B) is regular.*

A scale of methods based on power series are the Abel-methods, (A_λ). Suppose that $\lambda > -1$ and recall that E_n^λ is the binomial number $\binom{n+\lambda}{n}$. Consider the power series method based on the sequence $p_n = E_n^\lambda$. The fact that $p(x) = \sum_{k=0}^{\infty} E_k^\lambda x^k$ converges for $|x| < 1$ motivates the following definition.

Definition 2.20. *We write $s_n \to s$ (A_λ) if $\sigma_\lambda(x) = (1-x)^{\lambda+1} \sum_{k=0}^{\infty} E_k^\lambda s_k x^k$ converges for $|x| < 1$ and tends to s as x tends to 1^-.*

We note in the above definition that it is customary to use $\sigma_\lambda(x)$ for what, in our notation, is simply the $p_s(x)$ that corresponds to λ.

The case $\lambda = 0$ is the ordinary Abel method in which the well-known 'Limit Theorem' of Abel provides the regularity of the method.

In general, regularity follows easily from theorem 2.17 and the observation that $p(1) = \infty$.

Theorem 2.21. *For $\lambda > -1$, the method (A_λ) is regular.*

A third method of power series type is based on the series for the natural logarithm. Let $p_n = \frac{1}{n+1}$ for $n = 0, 1, 2, \cdots$. The power series method that results is called the logarithmic method (L).

Definition 2.22. *We write* $s_n \to s\ (L)$ *if* $L(x) := \dfrac{-x}{\ln(1-x)} \displaystyle\sum_{k=0}^{\infty} \frac{s_k}{k+1} x^k$ *converges for* $|x| < 1$ *and* $\lim\limits_{x \to 1^-} L(x) = s$.

Theorem 2.23. *The method* (L) *is regular.*

This follows immediately from theorem 2.17. The next result is easily obtained also.

Theorem 2.24. $(L) \supseteq (A_0)$.

Proof. Suppose $s_n \to s(A_0)$ and observe that $\sigma_0(x)$ and $L(x)$ have the same radius of convergence. Define the kernel $K(x,t)$ by

$$K(x,t) = \begin{cases} -\dfrac{1}{\ln(1-x)} \cdot \dfrac{1}{1-t} & \text{if } 0 < t < x < 1, \\ 0 & \text{otherwise.} \end{cases}$$

Then $L(x) = \int_0^1 K(x,t)\sigma_0(t)\,dt$ and the result follows by theorem 2.8 since the integral transform is regular.

The above result is true with (A_0) replaced by (A_λ); however the appropriate integral transform is less tidy and we omit it here. A proof of the general case that does not rely on an integral transform has been given by Borwein in [35].

2.4 Euler summability

We conclude this chapter by considering an important scale of triangular matrix methods. See also chapter 4.

Suppose $q > 0$ and define the matrix (c_{nk}) by

$$c_{nk} = \begin{cases} \dfrac{1}{(1+q)^n} \dbinom{n}{k} q^{n-k} & \text{if } 0 \le k \le n, \\ 0 & \text{otherwise.} \end{cases}$$

Then

$$\sum_{k=0}^{\infty} c_{nk} = \sum_{k=0}^{n} c_{nk}$$

$$= \frac{1}{(1+q)^n} \sum_{k=0}^{\infty} \binom{n}{k} q^{n-k} = 1.$$

Write t_n^q for the 'Euler mean' $\sum_{k=0}^{n} c_{nk} s_k$.

Definition 2.25. *For $q > 0$, if $\lim_{n\to\infty} t_n^q = s$ then we say that $\{s_n\}$ is Euler summable to s and write $s_n \to s$ (E, q).*

Theorem 2.26. *For $q > 0$ the (E, q) method is regular.*

Proof. By theorem 2.1, it suffices to show that $\lim_{n\to\infty} c_{nk} = 0$. But, for fixed k, we have

$$\lim_{n\to\infty} c_{nk} = \lim_{n\to\infty} \frac{\binom{n}{k} q^{n-k}}{(1+q)^n} = \lim_{n\to\infty} \frac{n^k}{q^k \Gamma(k+1)} \left(\frac{q}{1+q}\right)^n = 0.$$

Lemma 2.27. *If $s_n \to 0\,(E,q)$ then $s_n = \mathbf{o}\left((2q+1)^n\right)$ and the order estimate is best possible.*

For details of the proof of lemma 2.27 see Hardy **T**[17] or Peyerimhoff **T**[21].

Part of the relationship between Euler and Borel summability is given in the next result.

Theorem 2.28. $(B) \supseteq (E, q)$.

Proof. Suppose $s_n \to s$ (E, q). Then, by lemma 2.27, the Borel mean converges so that

$$\begin{aligned} e^{-x} \sum_{n=0}^{\infty} \frac{s_n}{n!} x^n &= e^{-(q+1)x} \sum_{n=0}^{\infty} \frac{(qx)^n}{n!} \sum_{n=0}^{\infty} \frac{s_n}{n!} x^n \\ &= e^{-(q+1)x} \sum_{n=0}^{\infty} \sum_{k=0}^{n} \frac{q^{n-k} s_k}{k!(n-k)!} x^n \\ &= e^{-(q+1)x} \sum_{n=0}^{\infty} t_n^q \frac{\left((q+1)x\right)^n}{n!}. \end{aligned}$$

The result now follows since (B) summability is regular.

3

Borel's methods of summability

3.1 Basic definitions

Although the basic definitions have been given in the previous chapter, we shall, for completeness, repeat them here. We shall also introduce some additional notation.

Suppose that $\{s_n\}$, $n = 0, 1, \ldots$ is a sequence of complex numbers generated as the sequence of partial sums of the infinite series $\sum_{n=0}^{\infty} a_n$. We define:

$$S(x) = e^{-x} \sum_{n=0}^{\infty} \frac{s_n x^n}{n!}, \tag{3.1}$$

$$A(x) = \int_0^x \left(e^{-t} \sum_{n=0}^{\infty} \frac{a_n t^n}{n!} \right) dt. \tag{3.2}$$

Definition 3.1. *If*

$$\lim_{x \to \infty} S(x) = \sigma, \tag{3.3}$$

then we say that the sequence $\{s_n\}$ *is* **summable by Borel's method** (B) *to* σ*, and write*

$$s_n \to \sigma\, (B), \tag{3.4}$$

it being assumed that the series concerned are convergent.

We also say that the infinite series $\sum_{n=0}^{\infty} a_n$ is summable to sum σ by Borel's method (B), and write $\sum_{n=0}^{\infty} a_n = \sigma\, (B)$.

Definition 3.2. *If*

$$\lim_{x\to\infty} A(x) = \sigma\,, \tag{3.5}$$

then we say that the sequence $\{s_n\}$ *is* **summable by Borel's method** (B') *to* σ*, and write*

$$s_n \to \sigma\,(B')\,. \tag{3.6}$$

We also say that the infinite series $\sum_{n=0}^{\infty} a_n$ is summable to sum σ by Borel's method (B'), and write $\sum_{n=0}^{\infty} a_n = \sigma\,(B')$.

In what follows, we shall use the sequence and the series versions interchangeably. Whenever we use one form it is to be understood that there is an equivalent form involving the other. Further, there are many situations in which the actual value of the number σ is not relevant. In such cases, we shall use one of the following forms:

$$\{s_n\} \in (B) \quad \text{or} \quad \sum_{n=0}^{\infty} a_n \in (B)\,;$$

$$\{s_n\} \in (B') \quad \text{or} \quad \sum_{n=0}^{\infty} a_n \in (B')\,.$$

It is thought that Borel had hoped to define two equivalent methods of summability. However, the two methods given above are not equivalent. This is shown in the next section. However, a result given in chapter 5 will indicate the 'error' in Borel's thinking. In fact, if he had used 'a_{n+1}' in (3.2) instead of 'a_n', he would have obtained two equivalent methods.

3.2 Basic properties of Borel's methods

3.2.1 Common properties

All the results stated in this subsection are true for Borel's method (B') although they are stated only for Borel's method (B).

Theorem 3.3. (see, for example, Hardy **T**[17], theorem 122 or theorem 2.17 above). *Borel's method* (B) *is regular; that is, if* $s_n \to \sigma$ *as* $n \to \infty$*, then* $s_n \to \sigma\ (B)$.

Theorem 3.4. (see, for example, Hardy **T**[17], theorem 127). *Borel's method* (B) *is linear: that is, if* $s_n \to \sigma\,(B)$*, if* $t_n \to \tau\,(B)$*, and if* λ *is any complex number, then* $(s_n + t_n) \to (\sigma + \tau)\,(B)$*, and* $(\lambda s_n) \to (\lambda\sigma)\,(B)$.

The proof of this is easy.

3.2.2 Interrelationships

In this section we give the interrelationships between the methods (B) and (B'). Here we see why Borel did not obtain two equivalent methods. Theorems 3.6, 3.7, and 3.5 illustrate this point.

Theorem 3.5. (see, for example, Hardy **T**[17], theorem 123). *Suppose that* $A'(x) \to 0$ *as* $x \to \infty$. *Then* $s_n \to \sigma\ (B)$ *if and only if* $s_n \to \sigma\ (B')$.

Theorem 3.6. (see, for example, Hardy **T**[17], theorem 124). *If* $s_n \to \sigma\ (B)$ *then* $s_n \to \sigma\ (B')$.

Theorem 3.7. (see, for example, Hardy **T**[17], theorem 125). *There is a sequence* $\{s_n\}$ *with the property that* $s_n \to \sigma\ (B')$ *and* s_n *is not summable* (B).

Note that

$$S(x) = A(x) + A'(x). \tag{3.7}$$

An example to illustrate theorem 3.7 must have $A'(x)$ not tending to zero. Hardy **T**[17] gives

$$a_n = \sum_{k=0}^{\infty} \frac{(-1)^k (2k+2)^n}{(2k+1)!},$$

from which it follows that

$$A'(x) = e^{-x} \sum_{k=0}^{\infty} \frac{(-1)^k}{(2k+1)!} e^{(2k+2)x} = \sin\left(e^x\right),$$

and so $\lim_{x\to\infty} A(x) = \int_0^{\infty} \sin\left(e^x\right) dx = \int_1^{\infty} \frac{\sin u}{u} du = \frac{\pi}{2}$.

However, we do have the following result of 'near equivalence':

Theorem 3.8. (see, for example, Hardy **T**[17], theorem 126). *The following two statements are equivalent:*

$$\sum_{n=0}^{\infty} a_n = s\,(B);$$

$$\sum_{n=1}^{\infty} a_n = (s - a_0)\,(B').$$

Theorem 3.9. (see, for example, Hardy **T**[17], theorem 127). *Borel's method (B) is left translative, that is,*

$$\textit{if} \sum_{n=0}^{\infty} a_n = \sigma\,(B), \ \textit{then} \sum_{n=1}^{\infty} a_n = (\sigma + a_0)\,(B).$$

Borel's method (B) is not right translative, that is,

$$\textit{if} \sum_{n=1}^{\infty} a_n = \sigma\,(B),$$

then it does not automatically follow that $\sum_{n=0}^{\infty} a_n = (\sigma - a_0)\,(B)$.

3.3 Extensions

3.3.1 Absolute summability

A summability method (T) is extended to *absolute summability* when the requirement of the convergence of the mean is replaced by a requirement of the mean being of *bounded variation.*

Definition 3.10. *If*

$$S(x) \in BV[0, \infty]\,, \tag{3.8}$$

then we say that the sequence $\{s_n\}$ *is* **absolutely summable by Borel's method** $|B|$, *and write* $\{s_n\} \in |B|$.

We also say that the infinite series $\sum_{n=0}^{\infty} a_n$ is summable by Borel's method $|B|$, and write $\sum_{n=0}^{\infty} a_n \in |B|$.

Definition 3.11. *If*

$$A(x) \in BV[0, \infty]\,, \tag{3.9}$$

then we say that the sequence $\{s_n\}$ *is* **absolutely summable by Borel's method** $|B'|$, *and write* $\{s_n\} \in |B'|$.

We also say that the infinite series $\sum_{n=0}^{\infty} a_n$ is summable by Borel's method $|B'|$, and write $\sum_{n=0}^{\infty} a_n \in |B'|$.

Note that (3.8) implies that $\lim_{x\to\infty} S(x)$ exists. Let the value of this limit be σ. Therefore, we shall write

$$s_n \to \sigma\,|B|,$$

or

$$\sum_{n=0}^{\infty} a_n = \sigma\,|B|.$$

Thus we have the following easy result:

Theorem 3.12. *If $s_n \to \sigma\,|B|$, then $s_n \to \sigma\,(B)$.*

Another way to express summability $|B'|$ is to note that it is equivalent to the absolute convergence of the integral $\int_0^\infty A(x)\,dx$, that is, that $\int_0^\infty |A(x)|\,dx$ is convergent.

A full discussion of absolute summability will occur in chapter 5.

3.3.2 Strong summability

Since the ideas of strong summability for Borel's methods were not introduced until after they had been generalized, we shall delay discussion of this topic to chapter 5.

3.3.3 Normal and regular summability

Suppose that p is any positive integer. Consider the series $\sum_{k=p}^{\infty} a_n$. Suppose that this series is summable (B) for every p. It then follows from theorem 3.6 that this series is summable (B') for every p, and conversely, from theorem 3.9. In this case, we shall say that $\sum_{n=0}^{\infty} a_n$ is *normally* summable by Borel's methods.

Theorem 3.13. (see, Hardy **T**[17], §8.6). *$\sum_{n=0}^{\infty} a_n$ is normally summable by Borel's methods if and only if*

$$e^{-x}a^{(p)}(x) := e^{-x}\sum_{n=0}^{\infty} a_{n+p}\frac{x^n}{n!} \to 0 \quad as \quad x \to \infty\,.$$

If $\int_0^\infty e^{-x}|a^{(p)}(x)|\,dx$ is convergent for every p, then we say that $\{s_n\}$ is *regularly* summable by Borel's methods. See Hardy **T**[17], §8.6.

These ideas, introduced in Hardy's book **T**[17], appear to have limited importance and no applications.

3.4 Relationships with other methods

3.4.1 Euler methods

There are several ways of introducing Euler's methods. We shall here, as in the previous chapter, adopt the sequence-to-sequence matrix transformation given by

$$c_{n,k} = \begin{cases} \dfrac{1}{(q+1)^{n+1}} \dbinom{n+1}{k+1} q^{n-k} & (k \le n), \\ 0 & (k > n). \end{cases}$$

We shall call $t_n^q = \sum_{k=0}^{n} c_{n,k}\, s_k$, the qth Euler transform of the sequence $\{s_n\}$. The methods are denoted by (E, q). We shall write $s_n \to \sigma\,(E, q)$ whenever the limit of the qth Euler transform is σ. The methods satisfy the following properties:

(E, q) is regular, linear, and translative.

If $s_n \to \sigma\,(E, q)$ and $q > q'$, then $s_n \to \sigma\,(E, q')$.

The rth Euler transform of the qth Euler transform of the sequence $\{s_n\}$ is the $(q + r + qr)$th Euler transform of the sequence $\{s_n\}$.

See Hardy's book **T**[17] §8.2, for these and other properties.

Borel's methods are stronger than Euler's methods. This relationship is given by:

Theorem 3.14. (see, for example, Hardy **T**[17], theorem 128, Knopp [151] and [152], or theorem 2.28 above). *If $s_n \to \sigma\,(E, q)$, then we have both $s_n \to \sigma\,(B)$ and $s_n \to \sigma\,(B')$.*

3.4.2 Cesàro and Abel methods

Borel's methods are not compatible with the Cesàro methods (C, k). This can be expressed best in the following way. Let $\mathcal{B}$ represent the set of all sequences that are summable (B), let $\mathcal{B}'$ represent the set of all sequences that are summable (B'), and let $\mathcal{C}$ represent the set of all sequences that are summable (C, k). Let $\mathcal{S}$ represent the set of all sequences. Then all of the following sets are nonempty:

$$\mathcal{B} \cap \mathcal{C}, \quad \{\mathcal{S} - \mathcal{B}\} \cap \mathcal{C}, \quad \mathcal{B} \cap \{\mathcal{S} - \mathcal{C}\}, \quad \{\mathcal{S} - \mathcal{B}\} \cap \{\mathcal{S} - \mathcal{C}\};$$

$$\mathcal{B}' \cap \mathcal{C}, \quad \{\mathcal{S} - \mathcal{B}'\} \cap \mathcal{C}, \quad \mathcal{B}' \cap \{\mathcal{S} - \mathcal{C}\}, \quad \{\mathcal{S} - \mathcal{B}'\} \cap \{\mathcal{S} - \mathcal{C}\};$$

Abel's method (A), which behaves, in many ways, like the Cesàro methods, has the same incompatibility relationships with Borel's methods. Fig. 3.1., adapted from Peyerimhoff **T**[21], illustrates this incompatibility by showing the comparable conditions on the sequence $\{s_n\}$:

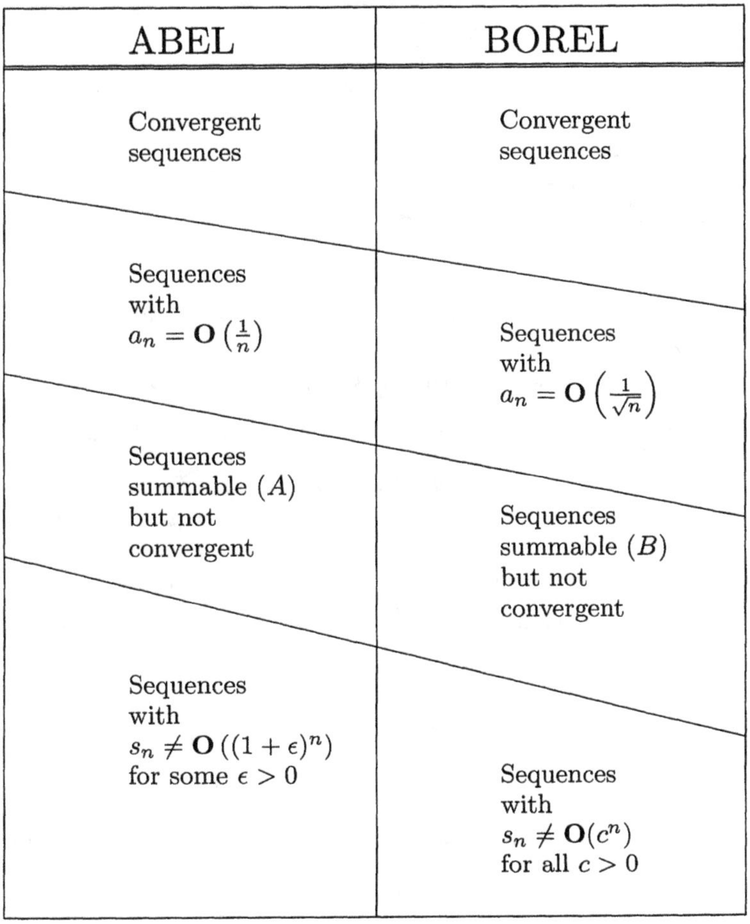

Further relationships with the Cesàro and Abel methods will appear in the section on Tauberian theorems below.

3.4.3 Other methods

In view of properties of the type that are stated in the previous section, it is clear that Borel's methods will only have simple relationships with those methods that are, in a sense, closely related to themselves (or to Euler's methods). See chapter 4 for some results.

3.5 Abelian theorems

Certain results are known as *Abelian* theorems because they are of the nature of Abel's theorem in the limit of a power series at its radius of convergence:

Theorem 3.15. *If* $\sum_{n=0}^{\infty} a_n\, x^n$ *has radius of convergence* ρ, *and if* $\sum_{n=0}^{\infty} a_n\, \rho^n$ *is convergent, then* $\lim_{x \to \rho^-} \sum_{n=0}^{\infty} a_n\, x^n = \sum_{n=0}^{\infty} a_n\, \rho^n$.

For Borel's methods, we have

Theorem 3.16. (see, for example, Hardy **T**[17], theorems 129 and 130). *If the power series* $\sum_{n=0}^{\infty} a_n\, z^n$ *is summable* (B') *at a point* P *in the complex plane, then it is summable at every point on the ray* OP. *Further, if* Q *is any point on the ray* OP, *then the series is uniformly summable on* QP.

Note. A power series $\sum a_n\, z^n$ is said to be *uniformly* summable for $0 \leq z \leq a$ if the appropriate summability mean is uniformly convergent on $0 \leq z \leq a$.

Theorem 3.17. (see, for example, Hardy **T**[17], theorem 132). *If the power series* $\sum_{n=0}^{\infty} a_n\, z^n$ *is summable* (B') *at a point* P *in the complex plane, then its* (B') *sum on* OP *is an analytic function of* z, *regular inside the circle with* OP *as diameter.*

This type of theorem is useful in finding analytic continuations of power series. See chapter 6 for more details.

3.6 Tauberian theorems

Results which give conditions under which a sequence, known to be summable by the method P, is also summable by the method Q, are known as *Tauberian* theorems. Although the standard texts such as Hardy **T**[17] and Peyerimhoff **T**[21] have sections on Tauberian theorems, the most comprehensive treatment is give by Pitt **T**[22].

As examples of theorems that are easy to apply, we have:

Theorem 3.18. (see, for example, Hardy **T**[17], theorem 143). *If* $\sum_{n=0}^{\infty} a_n = \sigma\,(B)$ *and* $a_n = \mathbf{o}(n^{-\frac{1}{2}})$, *then* $\sum_{n=0}^{\infty} a_n = \sigma$.

Theorem 3.19. (see, for example, Hardy **T**[17], theorem 147). *Suppose that* $\rho \geq -\frac{1}{2}$. *If* $\sum_{n=0}^{\infty} a_n = \sigma\,(B)$ *and* $a_n = \mathbf{o}(n^{\rho})$, *then*

$$\sum_{n=0}^{\infty} a_n = \sigma\,(C, 2\rho + 1).$$

However, Hardy uses theorem 3.18 as a stepping stone to obtain the stronger result:

Theorem 3.20. (see, for example, Hardy **T**[17], theorem 156). *If* $\sum_{n=0}^{\infty} a_n = \sigma\,(B)$ *and* $a_n = \mathbf{O}(n^{-\frac{1}{2}})$, *then* $\sum_{n=0}^{\infty} a_n = \sigma$.

The best Tauberian theorem for Borel's methods needs much more powerful machinery than the elementary techniques required for theorems 3.18 and 3.20. The theorem developed by Vijayaraghavan (see Pitt **T**[22] or Vijayaraghavan [301]) is required to obtain:

Theorem 3.21. (see, for example, Hardy **T**[17], theorem 241, or Pitt **T**[22], theorems 5 and 6). *If* $s_n \to \sigma\,(B)$ *and* $\liminf\,(s_n - s_m) \geq 0$ *as* $m \to \infty$, $n > m$, $\frac{n-m}{\sqrt{m}} \to 0$, *then* $s_n \to \sigma$ *as* $n \to \infty$.

4

Relations with the family of circle methods

In this chapter, we give a survey of the family of circle methods, as developed by Meyer–König in [185]. These methods are related to Borel's methods in that, under certain order conditions on the sequence $\{s_n\}$, they are essentially equivalent. We shall begin with Knopp's version of Euler's method, as first developed by Knopp in 1922 [151], [152]. We shall then survey the Taylor methods, defined by Hardy and Littlewood in 1916 [113]. Included in the survey is Valiron's method, given by Hardy and Littlewood in [113]. In order to assist in getting the relationships amongst these methods, Meyer–König introduced a new method in [185].

4.1 Euler–Knopp summability methods

The Euler–Knopp method. (E_p) $\quad 0 < p < \infty$.

Knopp's version of Euler's method is the positive, regular, triangular matrix:

$$E_p = (E_p)_{n,k} = \frac{1}{2^{np}} \binom{n}{k} (2^p - 1)^{n-k} \qquad (n, k \geq 0),$$

so that

$$E_p\, s = (E_p\, s)_n = \frac{1}{2^{np}} \sum_{k=0}^{n} \binom{n}{k} (2^p - 1)^{n-k}\, s_k\,.$$

If u, v are positive, then it is easy to see that

$$E_u\, E_v = E_{u+v}\,.$$

If $0 < p < q$, this gives that

$$E_q = E_{q-p}\, E_p\,,$$

so that

$$E_q\, s = E_{q-p}\,(E_p\, s)\,,$$

yielding that

$$s_n \to \sigma\,(E_p) \Rightarrow s_n \to \sigma\,(E_q)\,.$$

In this chapter, Borel's method is written in the form

$$B\, s = (B\, s)_{x,k} = B(x; s_k) = e^{-x} \sum_{k=0}^{\infty} \frac{x^k}{k!}\, s_k \qquad (x > 0)\,.$$

If $C = (C_{n,k})$ denotes a regular matrix, then

$$B\, C = (B\, C)_{n,k} = e^{-x} \sum_{j=0}^{\infty} \frac{x^j}{j!}\, C_{j,k} \qquad (x > 0; k = 0, 1, \ldots)\,,$$

giving

$$(B\, C)(x; s_k) = B(x; C\, s)\,.$$

With E_p in the place of C, this is

$$(B\, E_p\, s)_{x,k} = (B\, s)_{y,k} \qquad \text{with} \qquad y = \frac{x}{2^p} > 0\,,$$

giving

$$B(y; s_k) = B(x; E_p\, s) \qquad \text{with} \qquad y = \frac{x}{2^p} > 0\,,$$

so that

$$s_n \to \sigma\,(E_p) \Rightarrow s_n \to \sigma\,(B)\,.$$

Here we consider the set of z for which a summability method transforms the geometric series to a convergent series. If a positive, regular matrix C is applied to the geometric series $\sum_{n=0}^{\infty} z^n$, the result is

$$C\left(n; \frac{1 - z^{k+1}}{1 - z}\right) = C\left(n; \frac{1}{1 - z}\right) - \frac{z}{1 - z}\, C(n; z^k) \qquad (z \neq 1)\,.$$

The set of z for which $\sum_{n=0}^{\infty} z^n = \dfrac{1}{1 - z}\,(C)$ is a region, denoted by $\mathcal{G}'(C)$.

Since

$$E_p(n; z^k) = \frac{(2^p - 1 + z)^n}{2^{pn}}\,, \qquad B(x; z^k) = e^{-x(1-z)}\,,$$

it follows that

$$\mathcal{G}'(E_p) = \{z : |z + 2^p - 1| < 2^p\}, \qquad \mathcal{G}'(B) = \{\Re(z) < 1\}.$$

These illustrate the implications $E_p \Rightarrow E_q$ for $O < p < q$, and $E_p \Rightarrow B$ for $p > 0$.

4.2 T_α methods

4.2.1 Definitions

Unless stated to the contrary, assume that $0 < \alpha < 1$. The Taylor methods T_α are given by the positive, regular matrix

$$(T_\alpha)_{n,k} = (1-\alpha)^{n+1}\binom{k}{n}\alpha^{k-n} \qquad (n, k = 0, 1, 2, \ldots)$$

(see Hardy and Littlewood, [113]). Note that for $n = 0, 1, 2, \ldots,$

$$(1-\alpha)^{n+1}\sum_{k=n}^{\infty}\binom{k}{n}\alpha^{k-n} = (1-\alpha)^{n+1}\sum_{k=0}^{\infty}\binom{n+k}{k}\alpha^{k} = 1$$

and that

$$\begin{aligned} T_\alpha\, s = (T_\alpha\, s)_n &= (1-\alpha)^{n+1}\sum_{k=n}^{\infty}\binom{k}{n}\alpha^{k-n} s_k \\ &= (1-\alpha)^{n+1}\sum_{k=0}^{\infty}\binom{n+k}{k}\alpha^{k} s_{n+k}\,. \end{aligned}$$

Thus, T_α is applicable to $\{s_n\}$ if, for each $n \geq 0$, the series

$$\sum_{k=n}^{\infty}\binom{k}{n}\alpha^{k}\, s_k$$

converges. Therefore the following condition is necessary and sufficient:

$$s_k = \mathbf{O}\left(k^{-n}\alpha^{-k}\right) \quad \text{for each } n \geq 0 \text{ as } k \to \infty\,. \tag{4.1}$$

It follows that T_α applies to the sequence $|s| = \{|s_n|\}$ whenever it applies to s, and also in cases where the radius of convergence of the power series

$$\sum_{n=0}^{\infty} a_n\, x^n = (1-x)\sum_{n=0}^{\infty} s_n\, x^n$$

is greater than α.

Theorem 4.1. (see Meyer–König, [185], theorem 2). *Suppose that* $0 < \mu < 1/\alpha, \sigma$ *arbitrary,* $s_n \sim \sigma\mu^n$ *as* $b \to \infty$ *(so that* T_α *is applicable to* $\{s_n\}$*). Then*

$$T_\alpha s \sim \frac{(1-\alpha)\sigma}{1-\alpha\mu}\left(\frac{(1-\alpha)\mu}{1-\alpha\mu}\right)^n .$$

Applying the method T_α to the geometric series gives

$$T_\alpha(n; z^k) = \frac{1-\alpha}{1-\alpha z}\left(\frac{(1-\alpha)z}{1-\alpha z}\right)^n .$$

This enables us to find the region $\mathcal{G}'(T_\alpha)$ as follows:

$$\mathcal{G}'(T_\alpha) = \begin{cases} \left|z - \frac{\alpha}{2\alpha-1}\right| < \frac{1-\alpha}{1-2\alpha} & \text{for } 0 < \alpha \le \frac{1}{3}, \\ \left|z - \frac{\alpha}{2\alpha-1}\right| < \frac{1-\alpha}{1-2\alpha},\ |z| < \frac{1}{\alpha} & \text{for } \frac{1}{3} < \alpha \le \frac{1}{2}, \\ \Re(z) < 1,\ |z| < 2 & \text{for } \alpha = \frac{1}{2}, \\ \left|z - \frac{\alpha}{2\alpha-1}\right| > \frac{1-\alpha}{1-2\alpha},\ |z| < \frac{1}{\alpha} & \text{for } \frac{1}{2} < \alpha \le 1. \end{cases}$$

4.2.2 T_α on series

Theorem 4.2. (see Meyer–König, [185], theorem 3). T_α *is applicable to* $\{s_n\}$ *if and only if* T_α *is applicable to* $\{a_n\}$.
$s_n \to \sigma\,(T_\alpha)$ *is equivalent to*

$$\sum_{n=0}^{\infty}(1-\alpha)^n \sum_{k=0}^{\infty}\binom{n+k}{k}\alpha^k a_{n+k} = \sum_{n=0}^{\infty}\frac{T_\alpha(n; a_k)}{1-\alpha} = \sigma .$$

If T_α is applicable to $\{a_n\}$, then the series

$$f(z) = \sum_{k=0}^{\infty} a_k z^k, \quad \frac{1}{n!}f^{(n)}(z) = \sum_{k=0}^{\infty}\binom{n+k}{k}a_{n+k}z^k \quad (n = 0, 1, 2, \ldots)$$

have radius of convergence $r \ge \alpha$. Now, Abel's theorem gives

$$\frac{1}{n!}f^{(n)}(\alpha) = \frac{1}{(1-\alpha)^{n+1}}T_\alpha(n; a_k) .$$

Therefore the Taylor expansion of $f(z)$ about α

$$\mathcal{P}(z-\alpha)=\sum_{n=0}^{\infty}\frac{1}{n!}f^{(n)}(\alpha)(z-\alpha)^n$$

converges at the point $z=1$ to sum σ. Define the circular region

$$\mathcal{G}=\mathcal{G}(T_\alpha)=\{z:|z-\alpha|<1-\alpha\}.$$

Theorem 4.3. (see Meyer–König, [185], theorem 4). *Suppose that T_α is applicable to $\{a_n\}$, so that*

$$f(z)=\sum_{k=0}^{\infty}a_k\,z^k$$

is analytic for $|z|<\alpha$. If $f(z)$ is also analytic for $z=\alpha$, then

$$\sum_{n=0}^{\infty}a_n=\sigma\,(T_\alpha)\,,$$

if and only if $f(z)$ has an analytic continuation in the region $\mathcal{G}(T_\alpha)$, and the Taylor series $\mathcal{P}(z-\alpha)$ representing this continuation converges to σ at the point $z=1$, that is, when $z-\alpha=1-\alpha$.

4.2.3 T_α and T_β

Theorem 4.4. (see Meyer–König, [185], theorem 5, or Wais, **T**[29]). *Suppose that $0<\alpha<1, 0<\beta<1$, and $\gamma=\alpha+\beta-\alpha\beta$, so that $\max\{\alpha,\beta\}<\gamma<1$. Then*

$$T_\alpha\,T_\beta=T_\beta\,T_\alpha=T_\gamma\,.$$

Corollary 4.5. *Suppose that $0<\beta<\gamma<1, 0<\mu<(1-\beta)/(\gamma-\beta)$, that T_γ is applicable to $\{s_n\}$, so that T_β is also applicable to $\{s_n\}$, and that $T_\beta s\sim\sigma\,\mu^n$ and $n\to\infty$. Then*

$$T_\gamma s\sim\frac{(1-\gamma)\sigma}{1-\beta-\mu(\gamma-\beta)}\left(\frac{(1-\gamma)\mu}{1-\beta-\mu(\gamma-\beta)}\right)^n.$$

This leads to

Theorem 4.6. (see Meyer–König, [185], theorem 6). *Suppose that $0<\beta<\gamma<1$, that $s_n\to\sigma\,(T_\beta)$, and that T_γ is applicable to $\{s_n\}$. Then $s_n\to\sigma\,(T_\gamma)$.*

4.2.4 T_α and B

If $g(z) = \sum_{k=0}^{\infty} s_k z^k = \frac{1}{1-z} \sum_{k=0}^{\infty} a_k z^k$ converges for $|z| \le \alpha$, then $z = \alpha$ is a regular point. Also, for some $\rho > 0$,

$$g(z) = \sum_{k=0}^{\infty} \frac{g^{(k)}(\alpha)}{k!}(z-\alpha)^k \qquad \text{for } |z-\alpha| < \rho$$

is convergent. Also, for $n \ge 0$

$$\frac{1}{n!} g^{(n)}(z) = \sum_{k=0}^{\infty} \binom{n+k}{k} s_{n+k} z^k \qquad \text{for } |z| < \alpha\,,$$

and the series on the right converges for $z = \alpha$. So, by Abel's theorem

$$\frac{1}{n!} g^{(n)}(z) = \frac{t_n}{(1-\alpha)^{n+1}}$$

Therefore

$$g(z) = \frac{1}{1-\alpha} \sum_{k=0}^{\infty} \frac{t_k}{(1-\alpha)^k}(z-\alpha)^k \qquad \text{for } |z-\alpha| < \rho\,,$$

and in general

$$\frac{1}{n!} g^{(n)}(z) = \frac{1}{(1-\alpha)^{n+1}} \sum_{k=0}^{\infty} \binom{n+k}{k} \frac{t_{n+k}}{(1-\alpha)^k}(z-\alpha)^k \qquad \text{for } |z-\alpha| < \rho\,.$$

Observe that the power series on the right belongs to the Borel polygon, so that $z = 0$ lies in the interior of the Borel polygon and that the function is regular, not only in $|z-\alpha| < \rho$, but in $|z| < \alpha$.

We denote the Borel sum of a Borel summable series $\sum_{k=0}^{\infty} c_k$ by $\mathcal{B}\left(\sum_{k=0}^{\infty} c_k\right)$. Because $s_n = \frac{g^{(n)}(0)}{n!}$, we obtain

Theorem 4.7. (see Meyer–König, [185], theorem 7). *Suppose that T_α is applicable to $\{s_n\}$ and that $\sum_{k=0}^{\infty} a_k z^k$ is regular at $z = \alpha$. Then $(T_\alpha s)_n = t_n$ is defined and, for $n \ge 0$, it follows that*

$$s_n = \frac{1}{(1-\alpha)^{n+1}} \mathcal{B} \left(\sum_{k=0}^{\infty} \binom{n+k}{k} \left(\frac{-\alpha}{1-\alpha} \right)^k t_{n+k} \right)$$

where the right hand side is **Borel** *summable.*

This leads immediately to

Corollary 4.8. *Suppose that* $0 < \alpha < 1/3$, $s_n \to \sigma\,(T_\alpha)$, *and* $\sum_{k=0}^{\infty} a_k z^k$ *is regular at* $z = \alpha$. *Then* $s_n = \mathbf{o}((1-2\alpha)^{-n}) \quad as\, n \to \infty$.

And now we obtain a sharpening of theorem 4.6, to:

Theorem 4.9. (see Meyer–König, [185]). *Suppose that* $0 < \beta < 1/3$, $s_n \to \sigma\,(T_\alpha)$ *and that* $\sum_{k=0}^{\infty} a_k z^k$ *is regular at* $z = \beta$. *Then* $s_n \to \sigma\,(T_\gamma)$ *for every* γ *satisfying* $\beta < \gamma < 1 - 2\beta$.

4.2.5 Translativity

We now give the translative properties of T_α. Let $\{t_n\}$ be the T_α-transform of $s_0 = 0, s_1, s_2, \ldots$ and $\{t'_n\}$ be the T_α-transform of $s_1, s_2, \ldots$.

Theorem 4.10. (see Meyer–König, [185], theorem 8). *If* $t'_n \to \sigma$ *then* $t_n \to \sigma$.

If $1/2 < \alpha < 1$ *and if* $t_n \to \sigma$, *then* $t'_n \to \sigma$.

If $0 < \alpha \le 1/2$, *if* $\sum_{k=1}^{\infty} a_k z^k$ *is regular at* $z = \alpha$ *and if* $t_n \to \sigma$, *then* $t'_n \to \sigma$.

4.3 Meyer–König's S_α methods

4.3.1 Definition

In 1947, Meyer–König [185] defined a new method similar to the T_α method as follows:

$$S_\alpha = (S_\alpha)_{n,k} = (1-\alpha)^{n+1} \binom{n+k}{k} \alpha^k \qquad (n, k \ge 0)\,.$$

The similarity of the transformations

$$T_\alpha s = (1-\alpha)^{n+1} \sum_{k=0}^{\infty} \binom{n+k}{k} \alpha^k s_{n+k}\,,$$

$$S_\alpha s = (1-\alpha)^{n+1} \sum_{k=0}^{\infty} \binom{n+k}{k} \alpha^k s_k \,,$$

shows how the matrix S_α is composed of elements of the matrix T_α, transposed to the left.

Since, for fixed n, $\binom{n+k}{k} = \binom{n+k}{n} \sim \frac{k^n}{n!}$ as $n \to \infty$, showing that the following condition is necessary and sufficient for the applicability of the S_α method:

$$s_k = \mathbf{O}\left(k^{-n}\alpha^{-k}\right) \text{ for each } n \geq 0 \text{ as } k \to \infty \,.$$

This is the same as the applicability condition for T_α (see (4.1)). And this leads to

Theorem 4.11. (see Meyer–König, [185], theorem 9). *Suppose that $0 < \mu < 1/\alpha$, σ arbitrary, such that $s_n \sim \sigma\,\mu^n$ as $n \to \infty$, so that S_α is applicable to $\{s_n\}$. Then*

$$S_\alpha s \sim \frac{(1-\alpha)\sigma}{1-\alpha\mu} \left(\frac{1-\alpha}{1-\alpha\mu}\right)^n .$$

For the geometric series, since S_α is applicable to $\{z^n\}$, we have

$$S_\alpha(n; z^k) = \left(\frac{1-\alpha}{1-\alpha z}\right)^{n+1} ,$$

leading to

$$\mathcal{G}'(S_\alpha) = \{z : |z - 1/\alpha| > (1-\alpha)/\alpha, |z| < 1/\alpha\} \,.$$

Therefore we see that $\mathcal{G}'(S_\alpha)$ is larger than $\mathcal{G}'(T_\alpha)$. The union of all $\mathcal{G}'(S_\alpha)$ for $\alpha \in (0,1)$ is larger than the union of all $\mathcal{G}'(T_\alpha)$. The union of all $\mathcal{G}'(S_\alpha)$ for $\alpha \in (0,1)$ lies in the Borel half plane $\mathcal{G}'(B)$ whereas the union of all $\mathcal{G}'(T_\alpha)$ lies in the disk $\{z : |z| \leq 3\}$.

4.3.2 Translativity

The translativity of the S_α method is given by:

Theorem 4.12. (see Meyer–König, [185], theorem 10). *Suppose that $\{\theta_n\}$ is the S_α-transform of $s_0 = 0, s_1, s_2, \ldots$. and that $\{\theta'_n\}$ is the S_α-transform of $s_1, s_2, \ldots$. Then $\{\theta_n\}$ is summable S_α if and only if $\{\theta'_n\}$ is summable S_α.*

Here we get $\theta'_0 = \theta_0/\alpha$ and, for $n \geq 1$,

$$\theta'_n = \frac{1}{\alpha}\left(\theta_n - (1-\alpha)\theta_{n-1}\right) ;$$

4.3.3 S_α on series

Theorem 4.13. (see Meyer–König, [185], theorem 11). *The applicability of S_α to $\{s_n\}$ is equivalent to the applicability of S_α to $\{a_n\}$. In other words,*

$$s_n \to \sigma\,(S_\alpha)$$

is equivalent to

$$\alpha \sum_{n=0}^{\infty}(1-\alpha)^n \sum_{k=0}^{\infty}\binom{n+k}{k}\alpha^k a_k = \frac{\alpha}{1-\alpha}\sum_{n=0}^{\infty} S_\alpha(n; a_k) = \sigma\,.$$

For $0 < \beta < \gamma < 1$ we have that the intersection of $\mathcal{G}'(S_\beta)$ with the circle $\{z : |z|, 1/\gamma\}$ lies inside $\mathcal{G}'(S_\gamma)$.

4.3.4 S_α and S_β

An easy calculation gives

Theorem 4.14. (see Meyer–König, [185], theorem 12). *Suppose that $0 < \alpha < 1, 0 < \beta < 1$, and $\gamma = \dfrac{\beta}{1-\alpha(1-\beta)}$, so that $\beta < \gamma < 1$. Then $T_\alpha S_\beta = S_\gamma$.*

It now follows that

Corollary 4.15. *Suppose that $0 < \beta < \gamma < 1$, $0 < \mu < \gamma(1-\beta)/(\gamma-\beta)$ and that S_γ is applicable to $\{s_n\}$, so that S_β is applicable to $\{s_n\}$. Suppose that $S_\beta s \sim \sigma\,\mu^n$ as $n \to \infty$. Then*

$$S_\gamma s \sim \frac{\beta(1-\gamma)\sigma}{\gamma(1-\beta)-\mu(\gamma-\beta)}\left(\frac{\beta(1-\gamma)\mu}{\gamma(\beta)-\mu(\gamma-\beta)}\right)^n .$$

This now leads to

Theorem 4.16. (see Meyer–König, [185], theorem 13). *Suppose that $0 < \beta < \gamma < 1, s_n \to \sigma\,(S_\beta)$ and that S_γ is applicable to $\{s_n\}$. Then $s_n \to \sigma\,(S_\gamma)$.*

4.3.5 S_α and E_p

The S_α methods are a product of the Euler–Knopp method E_p with the Taylor method T_α as follows:

Theorem 4.17. (see Meyer–König, [185], theorem 14). *Suppose that* $0 < \alpha < 1$, *and* $\alpha = 1/2^p$ *so that* $p > 0$. *Then*

$$S_\alpha = E_p T_\alpha .$$

Corollary 4.18. *Suppose that* $\mu > 0$ *and* $s_n \sim \sigma \mu^n$ *as* $n \to \infty$. *Then*

$$E_p s \sim \sigma \left(\frac{2_p - 1 + \mu}{2^p} \right)^n .$$

Corollary 4.19. *Suppose that* $\mu > 0$, T_α *(and therefore* S_α*) is applicable to* $\{s_n\}$, *and that* $T_\alpha s \sim \sigma \mu^n$ *as* $n \to \infty$. *Then*

$$S_\alpha s \sim \sigma(1 - \alpha + \alpha\mu)^n .$$

Finally,

$$s_n = \frac{(-1)^n}{(1-\alpha)\alpha^n} \mathcal{B} \left(\sum_{k=0}^{\infty} \binom{n+k}{k} \sum_{j=0}^{n+k} \binom{n+k}{j} \frac{\theta_j}{(\alpha-1)^j} \right),$$

where we have a Borel sum on the right as before.

4.3.6 Function theoretic considerations

Theorem 4.20. (see Meyer–König, [185], theorem 16). *Suppose that* S_α *is applicable to* $\{s_n\}$ *and that* $f(z) = \sum_{k=1}^{\infty} a_k z^k$ *is regular at* $z = \alpha$. *Then the sequence of partial sums of the expansion of*

$$h(w) = f\left(\frac{\alpha}{1 - (1-\alpha)w} \right)$$

in a neighbourhood of $w = 0$ *in powers of* w, *is, at* $w = 1$, *identical with the sequence* $\{(S_\alpha s)_n\}$, *so that* $s_n \to \sigma\,(S_\alpha)$ *if and only if the above expansion converges to the value* 1 *at* $w = 1$.

This leads to

$$\mathcal{G}(S_\alpha) = \left\{ z : \left| z - \frac{1}{2-\alpha} \right| < \frac{1-\alpha}{2-\alpha} \right\}.$$

Set $\mu = \dfrac{1}{2-\alpha}$ so that $\frac{1}{2} < \mu < 1$. Then $\mathcal{G} = \mathcal{G}(S_\alpha)$ is the interior of a circle tangent internally to the circle $|z| = 1$ at $z = 1$. The point $z = \alpha$ lies in $\mathcal{G}$ and to the left of the centre $z = \mu$.

Denote by $\mathcal{L}$, the linear image of $\mathcal{G}$ to itself, such that the boundary points of $\mathcal{G}$ on the real axis are fixed points, and $z = \alpha$ is transformed on to $z = \mu$. Thus, the points

$$2\mu - 1 \qquad \alpha \qquad 1$$

are transformed to

$$2\mu - 1 \qquad \mu \qquad 1$$

The inverse $\mathcal{L}^{-1}$ of this is defined by the function

$$l^{(-1)}(z) = \frac{\alpha}{2-(2-\alpha)z}.$$

Suppose that S_α is applicable to $\{S_n\}$ and $f(z) = \sum_{k=0}^{\infty} a_k z^k$ is analytic at $z = \alpha$. The $\mathcal{L}$ is generated from $f(z)$ by

$$\overline{f}(z) = f\left(l^{(-1)}(z)\right)$$

being analytic at $z = \mu$.

If h is as defined in theorem 4.20, then in a neighbourhood of $y = 0$

$$\overline{f}(a + y) = h\left(\frac{y}{1-\mu}\right).$$

Theorem 4.21. (see Meyer–König, [185], theorem 17). *Suppose that S_α is applicable to $\{s_n\}$ and that $f(z) = \sum_{k=0}^{\infty} a_k z^k$ is regular at $z = \alpha$. Then $s_n \to \sigma\ (S_\alpha)$ if and only if $f(z)$ can be continued analytically into the region $\mathcal{G}(S_\alpha)$.*

Thus $\overline{f}(z)$ is analytic in $\mathcal{G}(S_\alpha)$. If it converges for $|y| < 1 - \mu$, then the power series expansion of $\overline{f}(\mu + y)$, in powers of y, converges at $y = 1 - \mu$ to σ.

4.4 Relations of T_α and S_α with E_p and B

$\mathcal{G}'(S_\alpha)$ and $\mathcal{G}'(B)$ have been defined above. They show that S_α and B are not comparable.

Suppose that $\sum_{k=0}^{\infty} s_k x^k$ has a positive radius of convergence. Thus $\sum_{k=0}^{\infty} \frac{s_k x^k}{k!}$ is convergent. Therefore the following series are convergent:

$$S_\alpha s = (1-\alpha)^{n+1} \sum_{k=0}^{\infty} \binom{n+k}{k} \alpha^k \, s_k \qquad \text{for } n \geq 0\,,$$

$$B\, s = b(x) = e^{-x} \sum_{k=0}^{\infty} \frac{x^k}{k!}\, s_k \qquad \text{for } x \geq 0\,.$$

The following relation:

$$S_\alpha s = \frac{w^{n+1}}{n!} \int_0^\infty x^n e^{-wx} b(x)\, dx \qquad \text{with} \quad w = \frac{1-\alpha}{\alpha}\,;$$

leads to

Theorem 4.22. (see Meyer–König, [185]). *Suppose that* $0 < \alpha < 1, -\infty < \mu < \frac{1-\alpha}{\alpha}$*, that* S_α *is applicable to* $\{s_n\}$ *and that* $b(x) \sim \sigma\, e^{\mu x}$ *as* $x \to \infty$*. Then*

$$S_\alpha s \sim \frac{(1-\alpha)s}{1-\alpha-\alpha\mu} \left(\frac{1-\alpha}{1-\alpha-\alpha\mu} \right)^n \qquad \textit{as } n \to \infty\,.$$

Theorem 4.23. (see Meyer–König, [185], theorem 18). *Suppose that* $s_n \to \sigma\,(B)$ *and that* S_α *is applicable to* $\{s_n\}$*. Then* $s_n \to \sigma\,(S_\alpha)$*.*

Theorem 4.24. (see Meyer–König, [185], theorem 19). *All the methods* $E_p\,(p > 0)$*,* B*,* $S_\alpha\,(0 < \alpha < 1)$*, and* $T_\beta\,(0 < \beta < 1)$ *are consistent with one another.*

$\mathcal{G}'(E_p)$ is contained in $\mathcal{G}'(S_\alpha)$ if $0 < \alpha \leq 1/(2^{p+1}-1)$, but they intersect if $1/(2^{p+1}-1) < \alpha < 1$. But the portion of $\mathcal{G}'(E_p)$ that contains the z values such that S_α is applicable to $\{z^n\}$ is always interior to $\mathcal{G}'(S_\alpha)$.

Theorem 4.25. (see Meyer–König, [185], theorem 20). *Suppose that* $0 < \alpha < 1, p > 0$ *and* $\beta = \alpha/(\alpha + 2^p(1-\alpha))$*, so that* $0 < \beta < \alpha$*. Then*

$$S_\alpha\, E_p = S_\beta\,.$$

Corollary 4.26. *Suppose that $0 < \beta < 1$, that $p > 0$, that $0 < \mu < (1-\beta+2^p\beta)/(2^p\beta)$, that S_β is applicable to $\{s_n\}$, and that $E_p s \sim \sigma\,\mu^n$ as $n \to \infty$. Then*

$$S_\beta s \sim \frac{(1-\beta)\sigma}{1-\beta+2^p\beta-2^p\beta\mu}\left(\frac{1-\beta}{1-\beta+2^p\beta-2^p\beta\mu}\right)^n .$$

Theorem 4.27. (see Meyer–König, [185], theorem 21).

If $\alpha(2^{p+1}-1) < 1$ and $s_n \to \sigma\,(E_p)$, then $s_n \to \sigma\,(S_\alpha)$.

If $\alpha(2^{p+1}-1) \geq 1$, then the methods E_p and S_α are incompatible.

If $s_n \to \sigma\,(E_p)$ and S_α is applicable to $\{s_n\}$, then $s_n \to \sigma\,(S_\alpha)$.

Theorem 4.28. (see Meyer–König, [185], theorem 22). *Suppose that $0 < \alpha < 1, p > 0$ and that β, q are defined by*

$$\beta = \frac{\alpha}{2^p(1-\alpha)+\alpha}, \qquad 2^q = 2^P(1-\alpha)+\alpha ,$$

so that $0 < \beta < \alpha$ and $0 < q < p$. Then

$$T_\alpha\, E_p = E_q\, T_\beta .$$

Corollary 4.29. *Suppose that $f(z) = \sum_{k=0}^{\infty} a_k z^k$ converges for $|z| < 1$ and that $\sum_{k=0}^{\infty} a_k = \sigma\,(E_p)$ for some $p > 0$. For q satisfying $0 < q < p$, so that*

$$\beta = \frac{2^p - 2^q}{2^q(2^p-1)} \in (0,1) ,$$

the series expansion of $f(\beta+y)$ in powers of y is, at $y = 1-\beta$, E_q-summable to σ.

4.5 Relations of E_p, B, and S_α with T_α

Hardy and Littlewood [113] have shown that B and $T_{1/2}$ are equivalent if $s_n = \mathbf{o}(\sqrt{n})$. In investigating further, it is instructive to look at the various $\mathcal{G}$'s associated with the family of methods.

$$\mathcal{G}'(B) = \{z : \Re(z) < 1\} \qquad \mathcal{G}'(T_{1/2}) = \{z : \Re(z) < 1, |z| < 2\}$$

$$\mathcal{G}'(E_p) = \mathcal{G}'(T_\alpha) \quad \text{for } \alpha = \frac{2^p - 1}{2^{p+1} - 1}, \; 0 < \alpha \leq \frac{1}{3}, \; 0 < p \leq 1 .$$

$$\mathcal{G}'(E_p) = \mathcal{G}'(T_\alpha) \quad \text{for } \alpha = \frac{2^p - 1}{2^{p+1} - 1}, \; \frac{1}{3} < \alpha < \frac{1}{2}, \; 1 < p < \infty .$$

$$\mathcal{G}'(S_\beta) = \mathcal{G}'(T_\alpha) \quad \text{for } \alpha = \frac{1}{1-\beta}, \; \frac{1}{2} < \alpha < 1, \; 0 < \beta < 1 .$$

This leads to

Theorem 4.30. (see Meyer–König, [185], theorem 23). *Suppose that* $s_n = \mathbf{o}(\sqrt{n})$, *so that* $\sum_{k=0}^{\infty} a_k z^k$ *is convergent for* $|z| < 1$. *Then*

(a) $\sum_{k=0}^{\infty} a_k = \sigma(E_p)$ *if and only if, when* $\alpha = \dfrac{2^p - 1}{2^{p+1} - 1}$, *the expansion of* $f(\alpha + y)$ *in powers of* y *converges at* $y = 1 - \alpha$ *to* σ,

(b) $\sum_{k=0}^{\infty} a_k = \sigma(B)$ *if and only if, when the expansion of* $F(\frac{1}{2} + y)$ *in powers of* y *converges at* $y = \frac{1}{2}$ *to* σ,

(c) $\sum_{k=0}^{\infty} a_k = \sigma(S_\beta)$ *for some* β *in* $(0, 1)$, *if and only if, at* $\gamma = \dfrac{1}{2-\beta}$, *the expansion of* $f(\gamma + y)$ *in powers of* y *converges at* $y = 1 - \gamma$ *to* σ.

Definition 4.31. Valiron's method. $\{s_n\}$ *is* F_μ *summable* [1] *to* σ *if, for some* $\mu > 0$

$$F_\mu s = \sqrt{\frac{\mu}{\pi n}} \sum_{k=-n}^{\infty} \exp\left(-\frac{\mu k^2}{n}\right) s_{n+k} \to \sigma$$

as $n \to \infty$, *provided that the series is convergent for all* $n > 0$.

If $s_n = \mathbf{o}(\sqrt{n})$ and $\mu = (1-\alpha)/2\alpha$, then $s_n \to \sigma(T_\alpha)$ if and only if $s_n \to \alpha(F_\mu)$.

[1] Hardy and Littlewood call this (E, μ) summable.

Definition 4.32. Condition $V = V(\lambda,\mu,\nu)$ for a matrix A. *Suppose that $A = (a_n, k)$ $n,k = 0,1,\ldots$, is regular and that there are positive constants λ,μ and ν such that: for the positive integers n and m with $m = \lambda n + \rho$ and $0 \le \rho < \max(1,\lambda)$, and for $-\nu, n < j < \nu n$ (j a whole number), and $m + j \ge 0$,*

$$a_{n,m+j} = \sqrt{\frac{\mu}{\pi m}} \exp\left(-\frac{\mu j^2}{m} + \frac{j}{m}\mathbf{O}(1) + \frac{j^3}{m^2}\mathbf{O}(1)\right)\left[1 + \mathbf{O}\left(\frac{1}{m}\right)\right],$$

where the '$\mathbf{O}$'-terms are uniformly in ρ and j as $n \to \infty$.

Further, for each positive $\tau < \lambda$ and each $t > \lambda$,

$$\sum_{0\le k\le \tau n} a_{n,k}s_k = \mathbf{o}(1), \quad \sum_{tn \le k} a_{n,k}s_k = \mathbf{o}(1) \qquad (n \to \infty),$$

whenever s_n satisfies the condition $s_n = \mathbf{O}(n^h)$ for some positive h.

The matrices E_p, S_β and T_α all satisfy this condition.

Lemma 4.33. (required in the proof of Theorem 4.30). *Suppose that the matrix A satisfies $V(\lambda,\mu,\nu)$ and that*

$$s_n = \mathbf{O}\left(n^{h+\frac{1}{2}}\right) \qquad \textit{for some } h \ge 0\,.$$

Then $t_n = (A\,s)_n = \mathbf{O}(n^k)$ if and only if $t'_n = (F_\mu\, s)_n = \mathbf{O}(n^k)$.
'$\mathbf{O}$' may be replaced by '$\mathbf{o}$'.

Take $h = 0$ and '$\mathbf{O}$' replaced by '$\mathbf{o}$' in the above. This means that $s_n = \mathbf{o}\left(\sqrt{n}\right)$, so that:

E_p is equivalent to F_μ for $\mu = \dfrac{2^{p-1}}{2^p - 1}$ $\quad\left(0 < p, \dfrac{1}{2} < \mu < \infty\right)$;

S_β is equivalent to F_μ for $\mu = \dfrac{1-\beta}{2}$ $\quad\left(0 < \beta < 1, 0 < \mu < \dfrac{1}{2}\right)$;

T_α is equivalent to F_μ for $\mu = \dfrac{1-\alpha}{2\alpha}$ $\quad(0 < \alpha < 1, 0 < \mu < \infty)$;

E_p is equivalent to T_α for $\alpha = \dfrac{2^p - 1}{2^{p+1} - 1}$ $\quad\left(0 < p, 0 < \alpha < \dfrac{1}{2}\right)$;

S_β is equivalent to T_α for $\alpha = \dfrac{1}{2-\beta}$ $\quad\left(0 < \beta < 1, \dfrac{1}{2} < \alpha < 1\right)$.

Let $0 < \alpha < 1$, $0 < \gamma \le \alpha$, $\delta = \dfrac{\alpha - \gamma}{1 - \gamma}$, and $s_n = \mathbf{o}\left(\sqrt{n}\right)$. Then $0 \le \delta < \alpha$ and $T_\alpha = T_\delta \, T_\gamma$. Therefore

$$T_\alpha s \to \sigma \qquad \text{is equivalent to} \qquad T_\delta(T_\gamma s) \to \sigma\,. \tag{4.2}$$

Also, $(T_\gamma s)_n = \mathbf{o}\left(\sqrt{n}\right)$, and so, in the case $0 < \alpha < \frac{1}{2}$,

$$2^p = \frac{1-\delta}{1-2\delta}, \text{ equivalently } \delta = \frac{2^p - 1}{2^{p+1}-1} \quad \left(0 < \delta < \frac{1}{2}, p > 0\right)$$

and

$$2^q = \frac{1-\delta}{1-2\delta}, \text{ equivalently } \delta = \frac{2^q - 1}{2^{q+1}-1} \quad \left(0 \le \delta < \frac{1}{2}, q \ge 0\right)$$

we get from (4.2)

$$E_p s \to \sigma \qquad \text{is equivalent to} \qquad E_q(T_\gamma s) \to \sigma$$

which is therefore true if

$$p > 0, \quad 0 < \gamma \le \frac{2^p - 1}{2^{p+1} - 1}, \quad 2^q = \frac{2^p}{1 + \gamma(2^{p+1} - 1)} \qquad (0 \le q < p)\,. \tag{4.3}$$

In the case $\alpha = \frac{1}{2}$ we get from (4.2)

$$s \to \sigma\,(B) \qquad \text{if and only if} \qquad s \to \sigma\,(E_q\, T_\gamma)\,,$$

which is therefore true if

$$0 < \gamma \le \frac{1}{2}, \; 2^q = \frac{1}{2\gamma} \qquad\qquad (0 \le q)\,. \tag{4.4}$$

Finally, in the case $\frac{1}{2} < \alpha < 1$, write

$$\beta = \frac{2\alpha - 1}{\alpha}, \text{ equivalently } \alpha = \frac{1}{2-\beta} \quad \left(\frac{1}{2} < \alpha < 1,\; 0 < \beta < 1\right)$$

and

$$\epsilon = \frac{2\delta - 1}{\delta}, \text{ equivalently } \delta = \frac{1}{2-\epsilon} \quad \left(\frac{1}{2} < \delta < 1,\; 0 < \epsilon < 1\right),$$

while, for $0 \le \delta < \frac{1}{2}$, the previous relationship between δ and q holds. Then we can deduce from (4.2)

$$s_n \to \sigma\,(S_\beta)$$

if and only if

$$\begin{cases} \text{(I)} \;\; s_n \to \sigma\,(E_q T_\gamma) \text{ when } 0 \le \delta < \frac{1}{2}\,, \\ \text{(II)} \;\; s_n \to \sigma\,(B\,T_\gamma) \;\text{ when } \delta = \frac{1}{2}\,, \\ \text{(III)} \; s_n \to \sigma\,(S_\epsilon T_\gamma) \;\text{ when } \frac{1}{2} < \delta < 1\,, \end{cases}$$

which is therefore true if

$$0 < \beta < 1, \begin{cases} \text{(I)} \;\; \frac{\beta}{2-\beta} < \gamma \le \frac{1}{2-\beta},\; 2^q = \frac{1-\beta}{\gamma(2-\beta)-\beta} \;\; (0 \le q)\,, \\ \text{(II)} \;\; \gamma = \frac{\beta}{2-\beta}\,, \\ \text{(III)} \; 0 < \gamma < \frac{\beta}{2-\beta},\; \epsilon = \frac{\beta-\gamma(2-\beta)}{1-\gamma(2-\beta)} \;\; (0 < \epsilon < \beta)\,. \end{cases} \tag{4.5}$$

All the above results come together as

Theorem 4.34. (see Meyer–König, [185], theorem 24). *Suppose that $s_n = \mathbf{o}\left(\sqrt{n}\right)$ so that the series $\sum_{k=0}^{\infty} a_k z^k$ is convergent for $|z| < 1$. Then:*

(a) *Let p, γ, q be chosen as in (4.3). Then $\sum_{k=0}^{\infty} a_k = \sigma\,(E_p)$ if and only if the expansion of $f(\gamma + y)$ in powers of y is E_q-summable to σ when $y = 1 - \gamma$.*

(b) *Let γ, q be chosen as in (4.4). Then $\sum_{k=0}^{\infty} a_k = \sigma\,(B)$ if and only if the expansion of $f(\gamma + y)$ in powers of y is E_q-summable to σ when $y = 1 - \gamma$.*

(c) *Let $\beta, \gamma, q, \epsilon$ be chosen as in (4.5). Then $\sum_{k=0}^{\infty} a_k = \sigma\,(S_\beta)$ if and only if the expansion of $f(\gamma + y)$ in powers of y is in case (I) E_q-summable, in case (II) B-summable, and in case (III) S_ϵ-summable, to σ when $y = 1 - \gamma$.*

When $q = 0$, we get theorem 4.30.

4.6 Equivalence of $E_p, B, S_\alpha, T_\alpha$ for bounded sequences

The restriction on $\{s_n\}$ is increased to $s_n = \mathbf{O}(1)$. This leads to the equivalence of the methods.

Theorem 4.35. (see Meyer–König, [185], theorem 25). *The methods E_p $(p > 0)$, B, $S_\beta(0 < \beta < 1)$, $T_\alpha(0 < \alpha < 1)$, and $F_\mu(\mu > 0)$ are equivalent for bounded sequences.*

Because of the theorems given above, it is sufficient to show that

Lemma 4.36. *All the methods $F_\mu(\mu > 0)$ are equivalent for bounded sequences.*

It is already true that if $0 < \mu_1 < \mu_2$ and if $s_n \to \sigma\,(T_{\mu_2})$, then $s_n \to \sigma\,(T_{\mu_1})$. To prove the converse requires use of the essence of the proof of Hardy and Littlewood's '**O**' Tauberian theorem for Borel's method B (see [113]), along with Knopp's proof of the similar theorem for Euler's method E_p (see Knopp [152]). These give

Lemma 4.37. *Suppose that $\{s_n\}$ is a bounded sequence and that $s_n \to \sigma\,(T_\mu)$ for some $\mu > 0$. Then, for $\nu = 1, 2, \ldots$, as $n \to \infty$,*

$$H_n^\nu := \frac{1}{\sqrt{n}} \sum_{k=-n}^{\infty} \left(\frac{k^2}{n}\right)^\nu \exp\left(-\frac{-\mu k^2}{n}\right) s_{n+k} \to 0\,.$$

4.7 Tauberian theorems

This section extends Hardy and Littlewood's major Tauberian theorem for Borel's method B:

Theorem 4.38. (see Hardy and Littlewood [113], theorem). *Suppose that $\sum_{n=0}^{\infty} a_n = \sigma\,(B)$ and $a_n = \mathbf{O}\left(\frac{1}{\sqrt{n}}\right)$. Then $\sum_{n=0}^{\infty} a_n = \sigma$;*

and Knopp's similar theorem for the Euler method E_p:

Theorem 4.39. (see Knopp, [152], theorem). *Suppose that $\sum_{n=0}^{\infty} a_n = \sigma\,(E_p)$ and $a_n = \mathbf{O}\left(\frac{1}{\sqrt{n}}\right)$. Then $\sum_{n=0}^{\infty} a_n = \sigma$.*

Here, we now have

Theorem 4.40. (see Meyer–König, [185]). *Suppose that* $\sum_{n=0}^{\infty} a_n = \sigma\left(T_\alpha\right)$ *and* $a_n = \mathbf{O}\left(\frac{1}{\sqrt{n}}\right)$. *Then* $\sum_{n=0}^{\infty} a_n = \sigma$.

These are all included in

Theorem 4.41. (see Meyer–König, [185], theorem 26). *Suppose that* $\sum_{n=0}^{\infty} a_n = \sigma\left(S_\alpha\right)$ *and* $a_n = \mathbf{O}\left(\frac{1}{\sqrt{n}}\right)$. *Then* $\sum_{n=0}^{\infty} a_n = \sigma$.

The proof needs

Theorem 4.42. (see Meyer–König, [185], theorem 27). *Suppose that* $h \geq 0$. *If* $S_\alpha s = \mathbf{O}\left(n^h\right)$ *and* $a_n = \mathbf{O}\left(n^{h-\frac{1}{2}}\right)$, *then* $s_n = \mathbf{O}\left(n^h\right)$.
The result remains true if '$\mathbf{O}$*' is replaced by '*$\mathbf{o}$*'.*

5

Generalizations of Borel's methods

5.1 First attempts

The first generalization of Borel's method (B') appears to have been given by Hardy [110], in 1934:

Definition 5.1. *A series* $\sum_{n=0}^{\infty} a_n$ *is said to be summable* (B, α) *to sum* σ *if the series*

$$a^{(\alpha)}(t) = \sum_{n=0}^{\infty} a_n \frac{t^{\alpha n}}{\Gamma(\alpha n + 1)}$$

is convergent for all $t \geq 0$*, and if the integral*

$$\int_0^{\infty} e^{-t} a^{(\alpha)}(t)\, dt = \lim_{R \to \infty} \int_0^R e^{-t} a^{(\alpha)}(t)\, dt$$

is convergent to σ*.*

Later, Hardy, for consistency, denoted this method by (B', α) (see **T**[17], §9.14). We shall, for the moment, adopt this notation. Observe that the method $(B', 1)$ is the method (B'). He further noted that his new method was more powerful than the method (B') if $\alpha > 1$, in that it could sum series that diverged more rapidly. For example, he cited

$$0! - 1! + 2! - 3! + \dots$$

as being summable (B', α) when $\alpha > 1$, but not when $\alpha \leq 1$.

Hardy **T**[17] also gave a generalization of Borel's method (B) as follows:

Definition 5.2. *A series* $\sum_{n=0}^{\infty} a_n$ *is said to be summable* (B, α) *to sum* σ *if the series*

$$e^{-x}\sum_{n=0}^{\infty}\left(\sum_{k=0}^{\infty}a_k\frac{\left(\frac{n}{\alpha}\right)^k}{k!}\right)\frac{x^n}{n!}\to\sigma \quad as \quad x\to\infty.$$

This latter method is quite complicated. It is not easy to understand the definition. Nor is it easy to make any use of it. It is not surprising that this method has not found any favour amongst subsequent authors. We shall not mention it again.

5.2 Mittag-Leffler's functions

In 1904, Mittag-Leffler [194] published a fifth note on generalizations of Laplace–Abel integrals. In it, he defined the function

$$E^{(\alpha)}(x)=\sum_{n=0}^{\infty}\frac{x^n}{\Gamma(\alpha n+1)},$$

and developed its properties. He showed *inter alia* that it is a generalization of the exponential function e^x in that it satisfies

$$E^{(\alpha)}(x)\sim\frac{e^x}{\alpha}\quad(\alpha>0)\quad \text{as}\quad x\to\infty.$$

This was extended in 1960 by Borwein [40], first to define a Mittag-Leffler-type function

$$E^{(\alpha,\beta)}(x)=\sum_{n=0}^{\infty}\frac{x^n}{\Gamma(\alpha n+\beta)},$$

and also to show that

$$\sum_{n=N}^{\infty}\frac{x^n}{\{\Gamma(an+b)\}^c(n+p)^{qn+r}}\sim\frac{K}{\alpha}(\kappa x)^{(\frac{1}{2}-\beta)/\alpha}e^{(\kappa x)^{1/\alpha}} \tag{5.1}$$

for $a>0$, $ac+q>0$ and $aN+b>0$, and where $\alpha=ac+q$, $\beta=c\left(b-\frac{1}{2}\right)+r$, and $\kappa=e^{-q}\alpha^{\alpha}a^{-ac}$, $K=e^{-pq}a^{r-\beta}\alpha^{\beta}(2\pi)^{\frac{1}{2}(1-c)}$.

This is the basic result that allows the generalizations of Borel's methods given in the next section.

5.3 Borel-type methods

5.3.1 Definitions

In defining Borel-type methods, the following assumptions are required: $\alpha > 0$, β is real, and N is a positive integer satisfying $\alpha N + \beta > 0$. These properties will be assumed hereafter.

Suppose that σ is a complex number. Define

$$s_n = \sum_{k=0}^{n} a_k, \; s_{-1} = 0, \; \sigma_N = \sigma - s_{N-1}.$$

Borel-type sums are defined as follows:

$$s_{\alpha,\beta}(x) = \sum_{n=N}^{\infty} \frac{s_n \, x^{\alpha n+\beta-1}}{\Gamma(\alpha n+\beta)}\,;$$

$$a_{\alpha,\beta}(x) = \sum_{n=N}^{\infty} \frac{a_n \, x^{\alpha n+\beta-1}}{\Gamma(\alpha n+\beta)}\,.$$

It is known that the convergence, for all $x \geq 0$, of either series, implies the convergence, for all $x \geq 0$, of the other (see Borwein [37]). This will be shown later in this chapter.

Using these sums, we are led to define Borel-type means as follows:

$$S_{\alpha,\beta}(x) = \alpha\, e^{-x}\, s_{\alpha,\beta}(x)\,,$$

$$A_{\alpha,\beta}(x) = \int_0^x e^{-t}\, a_{\alpha,\beta}(t)\, dt\,.$$

These are the (B, α, β) and (B', α, β) means of the sequence $\{s_n\}$ respectively. We may also say that these are the (B, α, β) and (B', α, β) means of the series $\sum_{n=0}^{\infty} a_n$ respectively.

For convenience, we also define the following mean, the (B, α, β) mean of the sequence $\{a_n\}$:

$$T_{\alpha,\beta}(x) = \alpha\, e^{-x}\, a_{\alpha,\beta}(x)\,.$$

Borel-type methods of summability are defined as follows:

Definition 5.3. Ordinary summability (see Borwein[40])

(i) *If* $S_{\alpha,\beta}(x) \to \sigma$ *as* $x \to \infty$, *then we say that* $s_n \to \sigma\,(B, \alpha, \beta)$.

(ii) *If* $A_{\alpha,\beta}(x) \to \sigma_N$ *as* $x \to \infty$, *then we say that* $s_n \to \sigma\,(B', \alpha, \beta)$.

Definition 5.4. Absolute summability (see Borwein and Shawyer [50])

(i) *If* $S_{\alpha,\beta}(x) \to \sigma$ *as* $x \to \infty$, *and* $S_{\alpha,\beta}(x)$ *is of bounded variation with respect to* x *in the range* $[0, \infty)$, *then we say that* $s_n \to \sigma\,|B, \alpha, \beta|$.

(ii) *If* $A_{\alpha,\beta}(x) \to \sigma_N$ *as* $x \to \infty$, *and* $A_{\alpha,\beta}(x)$ *is of bounded variation with respect to* x *in the range* $[0, \infty)$, *then we say that* $s_n \to \sigma\,|B', \alpha, \beta|$.

Definition 5.5. Strong summability (see Borwein and Shawyer [50])

(i) *If* $\int_0^x e^t\,|S_{\alpha,\beta-1}(t) - \sigma|\,dt = \mathbf{o}(e^x)$, *then we say that* $s_n \to \sigma\,[B, \alpha, \beta]$.

(ii) *If* $\int_0^x e^t\,|A_{\alpha,\beta-1}(t) - \sigma_N|\,dt = \mathbf{o}(e^x)$, *then we say that* $s_n \to \sigma\,[B', \alpha, \beta]$.

From these we see that the Borel-type method (B, α, β) so defined is equivalent to the integral function method associated with the integral function

$$E^{(\alpha,\beta)}(x) = \sum_{n=N}^{\infty} \frac{x^n}{\Gamma(\alpha n + \beta)}, \tag{5.2}$$

where $\alpha > 0$, β is real, and N is a positive integer satisfying $\alpha N + \beta > 0$.

Further, Borwein has shown that the integral function method associated with the integral function (see (5.2) above)

$$\sum_{n=N}^{\infty} \frac{z^n}{\{\Gamma(an+b)\}^c (n+p)^{qn+r}}$$

for $a > 0$, $ac + q > 0)$ and $aN + b > 0$, is equivalent to the method $(B, ac+q, bc+r-\frac{c}{2}+\frac{1}{2})$. For details, see Borwein [40].

Proposition. The following are equivalent:

(A) $\displaystyle\sum_{n=N}^{\infty} \frac{a_n x^{\alpha n+\beta-1}}{\Gamma(\alpha n+\beta)}$ *is convergent for all* $x \geq 0$,

(B) $\displaystyle\sum_{n=N}^{\infty} \frac{s_n x^{\alpha n+\beta-1}}{\Gamma(\alpha n+\beta)}$ *is convergent for all* $x \geq 0$.

Proof. First assume that (A) is true. Then, given $\epsilon > 0$, there exists a positive integer M such that for $n > M$,

$$|a_n| < \epsilon^n \Gamma(\alpha n+\beta) < \epsilon^{n+1}\Gamma(\alpha n+\alpha+\beta);$$

so that, for $n > M$,

$$|s_n| < \sum_{k=0}^{M} |a_k| + (n-M)\epsilon^n \Gamma(\alpha n+\beta).$$

Hence, for all sufficiently large n,

$$|s_n| < (2\epsilon)^n \Gamma(\alpha n+\beta),$$

and (B) follows.

Now assume (B) is true. Then since

$$\lim_{n\to\infty} \left\{ \frac{\Gamma(\alpha n+\alpha+\beta)}{\Gamma(\alpha n+\beta)} \right\}^{1/n} = 1,$$

we obtain that $\displaystyle\sum_{n=N}^{\infty} \frac{s_n x^{\alpha n+\alpha+\beta-1}}{\Gamma(\alpha n+\alpha+\beta)}$ is convergent for all $x \geq 0$. Further,

$$\sum_{n=N}^{\infty} \frac{a_n x^{\alpha n+\beta-1}}{\Gamma(\alpha n+\beta)} = \sum_{n=N}^{\infty} \frac{s_n x^{\alpha n+\beta-1}}{\Gamma(\alpha n+\beta)} - \sum_{n=N}^{\infty} \frac{s_n x^{\alpha n+\alpha+\beta-1}}{\Gamma(\alpha n+\alpha+\beta)}, \tag{5.3}$$

and (A) follows immediately.

The summability methods $(B,1,1)$ and $(B',1,1)$ are Borel's methods (B) and (B') defined in Chapter 3. The summability methods $(B,\alpha,1)$ and $(B',\alpha,1)$ are due to Włodarski [310] and Hardy (as mentioned at the beginning of this chapter). Some of the properties of these particular cases will be given in a subsequent section.

It is assumed henceforth that the series defining $s_{\alpha,\beta}(x)$ and $a_{\alpha,\beta}(x)$ are convergent for all $x \geq 0$, and since the actual choice of N is clearly unimportant, it is assumed that $\alpha N+\beta \geq 2$. Thus, all of the functions

$$s_{\alpha,\beta}(x)\,,\; s_{\alpha,\beta-1}(x)\,,\; s_{\alpha,\beta}(x)\,,\; s_{\alpha,\beta-1}(x),$$

are continuous for all $x \geq 0$.

We shall also assume, without any loss of generality, that

$$a_0 = a_1 = \ldots = a_{N-1} = 0,$$

so that $\sigma_N = \sigma$.

Because the functions $S_{\alpha,\beta}(x)$ and $A_{\alpha,\beta}(x)$ are absolutely continuous for $x \geq 0$, the definitions of absolute summability can be re-written as

Definition 5.6. Absolute summability:

(i) *If* $S_{\alpha,\beta}(x) \to \sigma$ *as* $x \to \infty$, *and* $\int_0^\infty |S'_{\alpha,\beta}(x)|\, dx < \infty$, *then we say that* $s_n \to \sigma\, |B, \alpha, \beta|$.

(ii) *If* $A_{\alpha,\beta}(x) \to \sigma_N$ *as* $x \to \infty$, *and* $\int_0^\infty |A'_{\alpha,\beta}(x)|\, dx < \infty$, *then we say that* $s_n \to \sigma\, |B', \alpha, \beta|$.

In what follows, we shall need to make use of summability, to limit 0, of the sequence $\{a_n\}$. Thus we note the following:

(a) $a_n \to 0\, (B, \alpha, \beta)$ means that $T_{\alpha,\beta}(x) \to 0$ as $x \to \infty$.

(b) $a_n \to 0\, |B, \alpha, \beta|$ means that $a_n \to 0\, (B, \alpha, \beta)$ and $\int_0^\infty |T'_{\alpha,\beta}(x)|\, dx < \infty$,

(c) $a_n \to 0\, [B, \alpha, \beta]$ means that $\int_0^x e^t\, |T_{\alpha,\beta-1}(t)|\, dt = \mathbf{o}(e^x)$.

5.3.2 Preliminaries

Here we note some results that are used throughout this chapter:

Given a function $f(x)$, continuous for $x \geq 0$, and a positive number δ, we write the Cesàro integral of $f(x)$ as follows:

$$f_0(x) = f(x)\,; \quad f_\delta(x) = \{\Gamma(\delta)\}^{-1} \int_0^x (x-t)^{\delta-1} f(t)\, dt.$$

I *If* $\delta > 0$ *and* $f(x) = s_{\alpha,\beta}(x)$, *then for* $x > 0$,

$$f_\delta(x) = s_{\alpha,\beta+\delta}(x) = \sum_{n=N}^{\infty} \frac{s_n x^{\alpha n+\beta+\delta-1}}{\Gamma(\alpha n+\beta+\delta)}.$$

II $a_{\alpha,\beta}(x) = s_{\alpha,\beta}(x) - s_{\alpha,\beta+\alpha}(x)$.

Proof. It follows from (5.3) that

$$a_{\alpha,\beta}(x) = s_{\alpha,\beta}(x)$$

$$- \sum_{n=0}^{\infty} \frac{s_n}{\Gamma(\alpha n+\beta)\Gamma(\alpha)} \int_0^x t^{\alpha n+\beta-1}(x-t)^{\alpha-1}\,dt$$

$$= s_{\alpha,\beta}(x) - \frac{1}{\Gamma(\alpha)} \int_0^x (x-t)^{\alpha-1}\,dt \sum_{n=0}^{\infty} \frac{s_n t^{\alpha n+\beta-1}}{\Gamma(\alpha n+\beta)\Gamma(\alpha)}$$

$$= s_{\alpha,\beta}(x) - s_{\alpha,\alpha+\beta}(x)\,,$$

the inversion being legitimate since

$$\int_0^x (x-t)^{\alpha-1}\,dt \sum_{n=0}^{\infty} \frac{s_n |t|^{\alpha n+\beta-1}}{\Gamma(\alpha n+\beta)\Gamma(\alpha)} < \infty\,.$$

III $\displaystyle\int_0^x e^{-t} a_{\alpha,\beta}(t)\,dt = \{\Gamma(\alpha)\}^{-1} \int_0^x e^{-u} s_{\alpha,\beta}(u)\,du \int_{x-u}^{\infty} t^{\alpha-1}e^{-t}\,dt.$

Proof. It follows from II that

$$\int_0^x e^{-t} a_{\alpha,\beta}(t)\,dt$$

$$= \int_0^x e^{-t} s_{\alpha,\beta}(t)\,dt - \frac{1}{\Gamma(\alpha)} \int_0^x e^{-t}\,dt \int_0^t (t-u)^{\alpha-1} s_{\alpha,\beta}(u)\,du$$

$$= \int_0^x e^{-t} s_{\alpha,\beta}(t)\,dt$$

$$- \frac{1}{\Gamma(\alpha)} \int_0^x e^{-u} s_{\alpha,\beta}(u)\,du \int_u^{\infty} (t-u)^{\alpha-1} e^{-(t-u)}\,dt$$

$$+ \frac{1}{\Gamma(\alpha)} \int_0^x e^{-u} s_{\alpha,\beta}(u)\,du \int_x^{\infty} (t-u)^{\alpha-1} e^{-(t-u)}\,dt$$

$$= \frac{1}{\Gamma(\alpha)} \int_0^x e^{-u} s_{\alpha,\beta}(u)\,du \int_{x-u}^{\infty} t^{\alpha-1}e^{-t}\,dt\,.$$

See Borwein [37] for these proofs in the case $\beta = 1$.

Henceforth, we shall write '$BV_x[0,\infty)$' to mean 'is of bounded variation with respect to x in the range $[0,\infty)$'.

5.3.3 Lemmas

Lemma 5.7.

$$\begin{array}{llcl}
\text{(i)} & A'_{\alpha,\beta}(x) & = & e^{-x}\, a_{\alpha,\beta}(x), \\
\text{(ii)} & A_{\alpha,\beta-1}(x) & = & e^{-x}\, a_{\alpha,\beta}(x) + \displaystyle\int_0^x e^{-t}\, a_{\alpha,\beta}(t)\, dt \\
 & & = & A_{\alpha,\beta}(x) + A'_{\alpha,\beta}(x), \\
\text{(iii)} & S'_{\alpha,\beta}(x) & = & \alpha\, e^{-x} \left\{ s_{\alpha,\beta-1}(x) - s_{\alpha,\beta}(x) \right\}, \\
\text{(iv)} & S_{\alpha,\beta-1}(x) & = & \alpha\, e^{-x}\, s_{\alpha,\beta-1}(x) = S_{\alpha,\beta}(x) + S'_{\alpha,\beta}(x), \\
\text{(v)} & T'_{\alpha,\beta}(x) & = & \alpha\, e^{-x} \left\{ a_{\alpha,\beta-1}(x) - a_{\alpha,\beta}(x) \right\}, \\
\text{(vi)} & T_{\alpha,\beta-1}(x) & = & \alpha\, e^{-x}\, a_{\alpha,\beta-1}(x) = T_{\alpha,\beta}(x) + T'_{\alpha,\beta}(x).
\end{array}$$

The proofs are very straightforward.

Lemma 5.8. *If* $\displaystyle\int_0^\infty |f(t)|\, dt < \infty$, *then* $\displaystyle\int_0^x e^x\, |f(t)|\, dt = \mathbf{o}(e^x)$ *as* $x \to \infty$.

Proof. Let $F(x) = \displaystyle\int_x^\infty |f(t)|\, dt$. Then

$$\int_0^x e^t\, |f(t)|\, dt = -F(0) + e^x\, F(x) - \int_0^x e^t F(t)\, dt = \mathbf{o}(e^x).$$

Lemma 5.9. *If* $f(x) \in BV_x[0,\infty)$ *and* $\displaystyle\int_0^\infty |\phi(t)|\, dt < \infty$, *then*

$$w(x) := \int_0^x f(x-t)\phi(t)\, dt \in BV_x[0,\infty).$$

Proof. Write $w(x) = \displaystyle\int_{-\infty}^\infty f^*(x-t)\phi^*(t)\, dt$, where

$$f^*(u) = \begin{cases} f(u) & (u \geq 0) \\ 0 & (u < 0) \end{cases}$$

$$\phi^*(u) = \begin{cases} \phi(u) & (u \geq 0) \\ 0 & (u < 0) \end{cases}$$

Then, for any partition $0 < x_0 < x_1 < \ldots < x_n$,

$$
\begin{aligned}
&\sum_{k=1}^{n} |w(x_k) - w(x_{k-1})| \\
&\le \int_{-\infty}^{\infty} \left\{ \sum_{k=1}^{n} |f^*(x_k - t) - f^*(x_{k-1} - t)| \right\} |\phi^*(t)|\, dt \\
&\le V_0^\infty f \int_0^\infty |\phi(t)|\, dt
\end{aligned}
$$

where $V_0^\infty f$ is the total variation of f in the range $[0, \infty)$. This completes the proof.
This is a special case of a lemma of Tatchell [274].

Lemma 5.10. *Suppose that $f(x)$ is continuous for $x \ge 0$ and that $\delta > 0$.*

(i) *If* $\lim_{x\to\infty} e^{-x} f(x) = \sigma$, *then* $\lim_{x\to\infty} e^{-x} f_\delta(x) = \sigma$.

(ii) *If* $\lim_{x\to\infty} \int_0^x e^{-t} f(t)\, dt = \sigma$, *then* $\lim_{x\to\infty} \int_0^x e^{-t} f_\delta(t)\, dt = \sigma$.

(iii) *If* $e^{-x} f(x) \in BV_x[0, \infty)$, *then* $e^{-x} f_\delta(x) \in BV_x[0, \infty)$.

(iv) *If* $\int_0^x e^{-t} f(t)\, dt \in BV_x[0, \infty)$, *then* $\int_0^x e^{-t} f_\delta(t)\, dt \in BV_x[0, \infty)$.

Proof. Let $F(x) = \int_0^x e^{-t} f(t)\, dt$. Then

$$\Gamma(\delta)\, e^{-x} f_\delta(x) = \int_0^x e^{-t} f(t) \phi(x - t)\, dt$$

and

$$\Gamma(\delta) \int_0^x e^{-t} f_\delta(t)\, dt = \int_0^x F(t) \phi(x - t)\, dt$$

where

$$\phi(u) = \Gamma(\delta)\, u^{\delta - 1}\, e^{-u}$$

is the kernel of a regular transformation.

Parts (i) and (ii) follow immediately. Parts (iii) and (iv) follow from lemma 5.9 above. See lemma 4 in Borwein and Shawyer [50] and lemma 2 in Borwein [37].

Lemma 5.11. *Suppose that $\delta > 0$. If $\int_0^x |f(t)|\,dt = \mathbf{o}(e^x)$, then*

$$\int_0^x |f_\delta(t)|\,dt = \mathbf{o}(e^x).$$

Proof. Let $g(t) = |f(t)|$. Thus $g_\delta(t) \geq |f_\delta(t)|$. By lemma 2(a) in Borwein [37], it follows that if $g_1(x) = \mathbf{o}(e^x)$, then we have that $g_{1+\delta}(x) = \mathbf{o}(e^x)$. Thus the result follows.

Lemma 5.12. (see Borwein [37], lemma 3). *If $\delta > 4, N$ is the integer such that $N \leq \delta/4 < N+1, w_n = e^{2\pi ni/\delta}, \gamma = 2\sin^2(\pi/\delta)$, and*

$$f(x) = E^{(\delta)}(x) = \sum_{n=0}^{\infty} \frac{x^n}{\Gamma(\delta n + 1)},$$

then

$$\text{(a)} \quad \delta\, f(x) = \sum_{n=-N}^{N} e^{xw_n} + \mathbf{O}(1),$$

$$\text{(b)} \quad \lim_{x\to\infty} \delta\, e^{-x} f(x) = 1,$$

$$\text{(c)} \quad e^{-x}\,\{f(x) - f_1(x)\} = \mathbf{O}(e^{-\gamma x}).$$

Proof. Results (a) and (b) are to be found in Hardy **T**[17], page 198. In view of (a), we have

$$f(x) - f_1(x) = \sum_{n=-N}^{N} (1 - w_{-n})e^{xw_n} + \mathbf{O}(1+x) = \mathbf{O}(e^{x\cos(2\pi/\delta)}),$$

and part (c) follows.

5.4 Relationships with respect to the parameter α

These relationships have not attracted much study. Various authors (Borwein [39], Good [96], Hardy **T**[17], and Włodarski [310] worked on the case where $\beta = 1$ before Borwein defined the general Borel-type methods (B, α, β) and (B', α, β). All of their results are subsumed by the following theorem.

Theorem 5.13. (see Borwein [40], result (I)). *If $\alpha > \gamma > 0$ and β, δ are real, if $s_n \to \sigma\,(B, \alpha, \beta)$, and if $\sum_{n=N}^{\infty} \dfrac{s_n x^{\gamma n+\delta-1}}{\Gamma(\gamma n+\delta)}$ is convergent for all $x \geq 0$, then $s_n \to \sigma\,(B, \gamma, \delta)$.*

The proof makes use of a general theorem due to Borwein , concerning inclusion relationships between integral function methods.

Borwein extended this theorem to absolute summability.

Theorem 5.14. (see Borwein [44], theorem). *If $\alpha > \gamma > 0$ and β, δ are real, if $s_n \to \sigma\,|B, \alpha, \beta|$, and if $\sum_{n=N}^{\infty} \dfrac{s_n x^{\gamma n+\delta-1}}{\Gamma(\gamma n+\delta)}$ is convergent for all $x \geq 0$, then $s_n \to \sigma\,|B, \gamma, \delta|$.*

These results also hold with B' replacing B.

5.5 Abelian relationships with respect to the parameter β

5.5.1 Interrelationships with same type

All of the following results are given by Borwein and Shawyer in [50]. Again, when there is a parallel result for B' methods, we shall only state the result for B methods.

Theorem 5.15. (see Borwein and Shawyer [50], theorem 2). *If $s_n \to \sigma\,|B, \alpha, \beta|$ then $s_n \to \sigma\,[B, \alpha, \beta]$.*

Proof. We are given both that $S_{\alpha,\beta}(x) - \sigma = \mathbf{o}(1)$ and that $\int_0^{\infty} \left|S'_{\alpha,\beta}(t)\right|\,dt < \infty$. Using lemmas 5.7 and 5.8, it follows that

$$\begin{aligned}
\int_0^x e^t\,|S_{\alpha,\beta-1}(t) - \sigma|\,dt &= \int_0^x e^t\,\left|S'_{\alpha,\beta}(t) + S_{\alpha,\beta}(t) - \sigma\right|\,dt \\
&= \int_0^x e^t\,\left|S'_{\alpha,\beta}(t) + \mathbf{o}(1)\right|\,dt \\
&= \mathbf{o}(e^x).
\end{aligned}$$

Theorem 5.16. (see Borwein and Shawyer [50], theorem 3). *If $s_n \to \sigma\,[B, \alpha, \beta]$ then $s_n \to \sigma\,(B, \alpha, \beta)$.*

Proof. Using relations in lemma 5.7, we have that

$$\int_0^x e^t \, |S_{\alpha,\beta-1}(t) - \sigma| \, dt = \mathbf{o}(e^x).$$

Hence it follows that

$$\begin{aligned}
|S_{\alpha,\beta} - \sigma| &= e^{-x} \, |e^x \, S_{\alpha,\beta}(x) - e^x \, \sigma| \\
&= e^{-x} \left| \int_0^x e^t \left\{ S_{\alpha,\beta}(t) + S'_{\alpha,\beta}(t) \right\} dt - \sigma e^x \right| \\
&= e^{-x} \left| \int_0^x e^t \left\{ S_{\alpha,\beta-1}(t) - \sigma \right\} dt - \sigma \right| \\
&\leq e^{-x} \int_0^x e^t \, |S_{\alpha,\beta-1}(t) - \sigma| + |\sigma| e^{-x} \\
&= \mathbf{o}(1).
\end{aligned}$$

Theorem 5.17. (see Borwein and Shawyer [50], theorem 5). *If* $s_n \to \sigma \, (B, \alpha, \beta)$ *then* $s_n \to \sigma \, [B, \alpha, \beta + 1]$.

The proof of this follows immediately from the definitions.

Theorem 5.18. (see Borwein [42], result II). *If* $s_n \to \sigma \, (B, \alpha, \beta)$ *then* $s_n \to \sigma \, (B, \alpha, \beta + \delta) \quad (\delta > 0)$.

Proof. From result I above, we see that

$$S_{\alpha,\beta+\delta}(x) = \{\Gamma(\delta)\}^{-1} \int_0^x (x - t)^{\delta - 1} e^{-(x-t)} S_{\alpha,\beta}(t) \, dt.$$

It is easy to check that the kernel $k(t) = \{\Gamma(\delta)\}^{-1} \, t^{\delta - 1} e^{-t}$ generates a regular integral transformation.

Theorem 5.19. (see Borwein and Shawyer [50], theorem 9). *If* $s_n \to \sigma \, [B, \alpha, \beta]$ *then* $s_n \to \sigma \, [B, \alpha, \beta + \delta] \quad (\delta > 0)$.

The proof of this is easily obtained using lemma 5.11.

Theorem 5.20. *If* $s_n \to \sigma \, |B, \alpha, \beta|$ *then* $s_n \to \sigma \, |B, \alpha, \beta + \delta| \quad (\delta > 0)$.

Although this result is not actually stated in Borwein and Shawyer [50], it is easily deduced from lemma 5.10 above.

Theorem 5.21. (see Borwein and Shawyer [50], theorem 11). *$s_n \to \sigma\,[B, \alpha, \beta]$ if and only if $s_n \to \sigma\,(B, \alpha, \beta)$ and*

$$\int_0^x e^t \left|S'_{\alpha,\beta}(t)\right| dt = \mathbf{o}(e^x).$$

Proof. From lemma 5.7, we obtain that

$$S_{\alpha,\beta-1}(t) - \sigma = S_{\alpha,\beta}(t) - \sigma + S'_{\alpha,\beta}(t),$$

whence

(*a*) $|S_{\alpha,\beta-1}(t) - \sigma| \le |S_{\alpha,\beta}(t) - \sigma| + |S'_{\alpha,\beta}|$, and

(*b*) $|S'_{\alpha,\beta}(t)| \le |S_{\alpha,\beta}(t) - \sigma| + |S_{\alpha,\beta-1} - \sigma|$.

Since $s_n \to \sigma\,[B, \alpha, \beta]$ implies that $s_n \to \sigma\,(B, \alpha, \beta)$, the use of (b) gives the proof of the necessity. The sufficiency follow easily from (a).

Note. This result indicates the following alternative definition for strong summability:

Definition 5.22. Strong summability:

(i) *If $S_{\alpha,\beta}(x) \to \sigma$ as $x \to \infty$, and*

$\displaystyle\int_0^\infty e^t |S'_{\alpha,\beta}(t)|\, dt = \mathbf{o}(e^x)$, *then we say that $s_n \to \sigma\,[B, \alpha, \beta]$.*

(ii) *If $A_{\alpha,\beta}(x) \to \sigma_N$ as $x \to \infty$, and*

$\displaystyle\int_0^\infty e^t |A'_{\alpha,\beta}(t)|\, dt = \mathbf{o}(e^x)$, *then we say that $s_n \to \sigma\,[B', \alpha, \beta]$.*

Definitions (5.6) and (5.22) indicate the extra conditions that must be imposed to proceed from ordinary (B, α, β) summability to the strong and absolute cases.

5.5.2 Interrelationships between types

Theorem 5.23. (see Borwein [42], theorem 1). *$s_n \to \sigma\,(B, \alpha, \beta)$ if and only if $s_n \to \sigma\,(B', \alpha, \beta)$ and $a_n \to 0\,(B, \alpha, \beta)$.*

Proof. Suppose that $s_n \to \sigma\,(B, \alpha, \beta)$. From theorem 5.18, we have that $s_n \to \sigma\,(B, \alpha, \beta + \alpha)$. Using II, we now obtain that $a_n \to \sigma\,(B, \alpha, \beta)$.

Further, III indicates that $A_{\alpha,\beta}(x)$ is obtained as a regular integral transformation of $S_{\alpha,\beta}(x)$, so that we obtain that $s_n \to \sigma\,(B', \alpha, \beta)$.

Suppose now that $s_n \to \sigma\,(B', \alpha, \beta)$ and $a_n \to 0\,(B, \alpha, \beta)$; that is,

$$A_{\alpha,\beta}(x) = \int_0^x e^{-t} a_{\alpha,\beta}(t), dt \to \sigma\,, \tag{5.4}$$

$$T_{\alpha,\beta}(x) = \alpha e^{-x} a_{\alpha,\beta}(x) \to 0\,. \tag{5.5}$$

Suppose that k is an integer and that $\delta = k\alpha > 4$. Let

$$\begin{aligned} b(x) &:= \sum_{r=0}^{k-1} a_{\alpha,\beta+\alpha r}(x)\,, \\ B(x) &:= \int_0^x e^{-t} b(t)\,dt\,, \\ f(x) &= E^{(\delta)}(x)\,, \\ \phi(x) &= \frac{d}{dx}\left\{e^{-x} f_{\delta-1}(x)\right\}\,. \end{aligned}$$

It now follows from (5.4) and lemma 5.10, that

$$\lim_{x\to\infty} B(x) = k\sigma\,; \tag{5.6}$$

and further, using II, that $b(x) = s_{\alpha,\beta}(x) - s_{\alpha,\beta+\delta}(x)$. Also

$$\begin{aligned} f_{\delta-1}(x) &= \sum_{n=1}^{\infty} \frac{1}{\Gamma(\delta n - \delta + 1)\Gamma(\delta - 1)} \int_0^x (x-t)^{\delta-2} t^{\delta n - \delta}\,dt \\ &= \sum_{n=1}^{\infty} \frac{x^{\delta n - 1}}{\Gamma(\delta n)}\,. \end{aligned}$$

It is easy to check that $\lim\limits_{m\to\infty} s_{\delta m}(x) = 0$, and so

$$\begin{aligned} s_{\alpha,\beta+\delta}(x) &= \sum_{n=1}^{\infty} \left\{s_{\alpha,\beta+\delta n}(x) - s_{\alpha,\beta+\delta n+\delta}(x)\right\} \\ &= \sum_{n=1}^{\infty} b_{\delta n}(x) \end{aligned}$$

$$= \sum_{n=1}^{\infty} \frac{1}{\Gamma(\delta n)} \int_0^x t^{\delta n - 1} b(x-t)\, dt$$

$$= \int_0^x b(x-t) f_{\delta-1}(t)\, dt,$$

the inversion of integration and sum being valid since

$$\int_0^x |b(x-t)| f_{\delta-1}(t)\, dt < \infty .$$

Integration by parts now yields that

$$e^{-x} s_{\alpha,\beta+\delta}(x) = \int_0^x B(x-t)\phi(t)\, dt . \tag{5.7}$$

We shall now show that $\phi(x)$ is the kernel of a regular transformation.

By lemma 5.12, we obtain that $\lim_{x\to\infty} \delta e^{-x} f(x) = 1$. Hence, by lemma 5.10

$$\lim_{x\to\infty} \int_0^x \phi(t)\, dt = \lim_{x\to\infty} e^{-x} f_{\delta-1}(x) = 1 ,$$

and

$$\lim_{x\to\infty} \phi(x) = \lim_{x\to\infty} e^{-x} \{ f_{\delta-2}(x) - f_{\delta-1}(x) \} = 0 .$$

Further, $\phi(x)$ is continuous for $x \geq 0$, and using lemma 5.12, we obtain that

$$\int_0^\infty |\phi(x)|\, dx \leq \frac{1}{\Gamma(\delta-2)} \int_0^\infty e^{-x}\, dx \int_0^x (x-t)^{\delta-3} |f(t) - f_1(t)|\, dt$$

$$= \int_0^\infty e^{-t} |f(t) - f_1(t)|\, dt < \infty .$$

It now follows from (5.6) and (5.7) that we have a regular transformation, so that

$$\lim_{x\to\infty} \alpha e^{-x} s_{\alpha,\beta+\delta}(x) = \sigma ;$$

so that, in fact, $s_n \to \sigma\,(B, \alpha, \beta+\delta)$.

To complete the proof, we need only note that use of II, lemma 5.10, and (5.5) gives

$$\lim_{x\to\infty} e^{-x} \{ s_{\alpha,\beta}(x) - s_{\alpha,\beta+\delta}(x) \} = 0 ;$$

so that

$$\lim_{x\to\infty} \alpha e^{-x} s_{\alpha,\beta}(x) = \sigma\,.$$

This completes the proof.

Theorem 5.24. (see Borwein and Shawyer [50], theorem 15). *$s_n \to \sigma\,[B,\alpha,\beta]$ if and only if $s_n \to \sigma\,[B',\alpha,\beta]$ and $a_n \to 0\,[B,\alpha,\beta]$.*

Proof. In view of theorem 5.21 it is sufficient to prove that $\int_0^x e^t|S'_{\alpha,\beta}(t)|\,dt = \mathbf{o}(e^x)$ *if and only if* $\int_0^x e^t|A'_{\alpha,\beta}(t)|\,dt = \mathbf{o}(e^x)$ *and* $\int_0^x e^t|T'_{\alpha,\beta}(t)|\,dt = \mathbf{o}(e^x)$.

Suppose first that

$$\int_0^x e^t|S'_{\alpha,\beta}(t)|\,dt = \mathbf{o}(e^x)\,;$$

that is, that

$$\int_0^x |s_{\alpha,\beta}(t) - s_{\alpha,\beta-1}(t)|\,dt = \mathbf{o}(e^x)\,.$$

In view of lemma 5.11, it follows that

$$\int_0^x |s_{\alpha,\beta+\alpha}(t) - s_{\alpha,\beta+\alpha-1}(t)|\,dt = \mathbf{o}(e^x)\,.$$

Thus

$$\begin{aligned}\int_0^x & |a_{\alpha,\beta}(t) - a_{\alpha,\beta-1}(t)|\,dt \\ &\le \int_0^x |s_{\alpha,\beta}(t) - s_{\alpha,\beta-1}(t)|\,dt \int_0^x |s_{\alpha,\beta+\alpha}(t) - s_{\alpha,\beta+\alpha-1}(t)|\,dt \\ &= \mathbf{o}(e^x)\,;\end{aligned}$$

so that

$$\int_0^x e^t|T'_{\alpha,\beta}(t)|\,dt = \mathbf{o}(e^x)\,.$$

Further, since

$$A'_{\alpha,\beta}(t) = T_{\alpha,\beta}(t) = T_{\alpha,\beta-1}(t) - T_{\alpha,\beta}(t)\,,$$

it follows that

$$\int_0^x e^t|A'_{\alpha,\beta}(t)|\,dt = \mathbf{o}(e^x)\,.$$

For the second part, we use the same notations as in the previous two theorems. Suppose now that $s_n \to \sigma\,[B,\alpha,\beta]$. This implies that $s_n \to \sigma\,(B,\alpha,\beta)$, and further that $s_n \to \sigma\,(B,\alpha,\beta+\nu)$ for every $\nu \geq 0$, so that we have

$$s_{\alpha,\beta+\delta}(x) \;-\; s_{\alpha,\beta+\delta-1}(x) = \mathbf{o}(e^x)\,. \tag{5.8}$$

Since we have both

$$\int_0^x e^t|A'_{\alpha,\beta}(t)|\,dt = \int_0^x |a_{\alpha,\beta}(t)|\,dt = \mathbf{o}(e^x)\,,$$

and

$$\int_0^x e^t|A'_{\alpha,\beta}(t)|\,dt = \int_0^x \alpha|a_{\alpha,\beta}(t) - a_{\alpha,\beta-1}|\,dt = \mathbf{o}(e^x)\,,$$

it follows that

$$\int_0^x |a_{\alpha,\beta}(t)|\,dt = \mathbf{o}(e^x)\,.$$

It now follows from lemma 5.11 that we have both

$$\int_0^x |s_{\alpha,\beta-1}(t) \;-\; s_{\alpha,\beta+\delta-1}(t)|\,dt = \mathbf{o}(e^x)\,,$$

and

$$\int_0^x |s_{\alpha,\beta}(t) \;-\; s_{\alpha,\beta+\delta}(t)|\,dt = \mathbf{o}(e^x)\,.$$

From these two estimates, in conjunction with (5.8), it follows that

$$\int_0^x |s_{\alpha,\beta}(t) \;-\; s_{\alpha,\beta-1}(t)|\,dt = \int_0^x e^t|S'_{\alpha,\beta}(t)|\,dt = \mathbf{o}(e^x)\,.$$

The proof is now complete.

Theorem 5.25. (see Borwein and Shawyer [50], theorem 14).
$s_n \to \sigma\,|B,\alpha,\beta|$ if and only if $s_n \to \sigma\,|B',\alpha,\beta|$ and $a_n \to 0\,|B,\alpha,\beta|$.

Proof. The ordinary summability part of this is given by theorem 5.23. It remains to deal with the bounded variation conditions.

Suppose that $S_{\alpha,\beta}(x) \in BV_x[0,\infty)$. From lemma 5.10 (iii), since $\alpha > 0$, it follows that $S_{\alpha,\alpha+\beta}(x) \in BV_x[0,\infty)$, and further, from II, that $T_{\alpha,\beta}(x) \in BV_x[0,\infty)$. Thus $a_n \to 0\,|B,\alpha,\beta|$.

Also,

$$A_{\alpha,\beta}(x) = \int_0^x e^{-t} a_{\alpha,\beta}(t)\, dt = \int_0^x e^{-u} s_{\alpha,\beta}(u)\phi(u)\, du\,,$$

where $\phi(u) = \int_v^\infty e^{-t} t^{\alpha-1}\, dt.$

Since $\int_v^\infty \phi(v)\, dv = \Gamma(\alpha+1) < \infty$, it now follows from lemma 5.10 (iv), that $A_{\alpha,\beta}(x) \in BV_x[0,\infty)$, so that $s_n \to \sigma\,|B',\alpha,\beta|$.

Suppose now that $A_{\alpha,\beta}(x) \in BV_x[0,\infty)$ and $T_{\alpha,\beta}(x) \in BV_x[0,\infty)$. Using the notation of the second part of the proof of theorem 5.23, we obtain that

$$e^{-x} a_{\alpha,\beta+\delta}(x) = \int_0^x B(x-t)\phi(t)\, dt$$

where k, δ, $b(x)$, $B(x)$, and $\phi(x)$ are as above. Note that

$$B(x) = \int_0^x e^{-t}\left\{s_{\alpha,\beta}(t) - s_{\alpha,\beta+\delta}(t)\right\} dt\,.$$

Since $A_{\alpha,\beta}(x) \in BV_x[0,\infty)$, we obtain that

$$\int_0^x e^{-t}\left\{s_{\alpha,\beta}(t) - s_{\alpha,\beta+\alpha}(t)\right\} dt \in BV_x[0,\infty)\,.$$

In view of lemma 5.10 (iv), we see that

$$\int_0^x e^{-t}\left\{s_{\alpha,\beta}(t) - s_{\alpha,\beta+\delta}(t)\right\} dt \in BV_x[0,\infty)\,,$$

so that $B(x-t)$ is of bounded variation with respect to x in the range $[t,\infty)$ uniformly for $t \geq 0$.

Thus, in view of lemma 5.9, it follows that $S_{\alpha,\beta+\delta}(x) \in BV_x[0,\infty)$, and further, from lemma 5.10 (ii), that $S_{\alpha,\beta}(x) - S_{\alpha,\beta+\delta}(x) \in BV_x[0,\infty)$, which is sufficient for the conclusion.

Theorem 5.26. (see Borwein [42], theorem 2). *$s_n \to \sigma\,(B',\alpha,\beta)$ if and only if $s_n \to \sigma\,(B,\alpha,\beta+1)$.*

Proof. We again use the notation developed in theorem 5.23. Suppose that $s_n \to \sigma\,(B',\alpha,\beta)$, so that

$$\lim_{x\to\infty} \int_0^x e^{-t} a_{\alpha,\beta}(t)\, dt = \sigma\,.$$

Again let $\delta = k\alpha > 4$ so that, as before, we have, for $x \geq 0$,

$$b(x) = \sum_{r=0}^{k-1} a_{\alpha,\beta+\alpha r}(x) = s_{\alpha,\beta}(x) - s_{\alpha,\beta+\delta}(x)\,.$$

Arguing as in theorem 5.23, we deduce that

$$\lim_{x\to\infty} \alpha e^{-x} s_{\alpha,\beta+\delta} = \sigma\,;$$

and further, using lemma 5.10 (i), we have that

$$\lim_{x\to\infty} \alpha e^{-x} s_{\alpha,\beta+\delta+1} = \sigma\,.$$

Further, for $x > 0$,

$$\int_0^x e^{-t} a_{\alpha,\beta}(t)\, dt = -\, e^{-x} a_{\alpha,\beta+1}(x) + \int_0^x e^{-t} a_{\alpha,\beta+1}(t)\, dt\,, \tag{5.9}$$

so that from lemma 5.10 (ii), we further obtain that

$$\lim_{x\to\infty} e^{-x} a_{\alpha,\beta+1}(x) = 0\,. \tag{5.10}$$

It now follows from lemma 5.10 (i) that

$$\lim_{x\to\infty} \alpha e^{-x} \left\{ s_{\alpha,\beta+1}(x) - s_{\alpha,\beta+\delta+1}(x) \right\} = 0\,,$$

so that

$$\lim_{x\to\infty} \alpha e^{-x} s_{\alpha,\beta+1}(x) = \sigma\,;$$

that is, $s_n \to \sigma\,(B, \alpha, \beta+1)$.

Suppose now that $s_n \to \sigma\,(B, \alpha, \beta+1)$. We obtain from theorem 5.26 that $a_n \to \sigma\,(B, \alpha, \beta+1)$, so that

$$\lim_{x\to\infty} e^{-x} a_{\alpha,\beta+1} = 0.$$

Using lemma 5.10 (i), we see that

$$\lim_{x\to\infty}\int_0^x e^{-t}a_{\alpha,\beta+1}(t)\,dt = 0\,.$$

Combining these two in (5.9) gives the result.

Theorem 5.26 explains why Borel did not obtain two equivalent methods when he introduced the (B) and the (B') methods. He was 'one out'!

Theorem 5.27. (see Borwein and Shawyer [50], theorem 18). *$s_n \to \sigma\,[B',\alpha,\beta]$ if and only if $s_n \to \sigma\,[B,\alpha,\beta+1]$.*

Proof. The ordinary summability part of this theorem is given by theorem 5.26. It remains to consider the order conditions.

If we are given that $s_n \to \sigma\,[B',\alpha,\beta]$, we have that

$$\int_0^x e^t|A'_{\alpha,\beta}(t)|\,dt = \alpha\int_0^x |a_{\alpha,\beta}(t)|\,dt = \mathbf{o}(e^x)\,.$$

From lemma 5.11, we obtain that

$$\int_0^x |a_{\alpha,\beta+1}(t)|\,dt = \mathbf{o}(e^x)\,.$$

and hence that

$$\int_0^x |a_{\alpha,\beta+1}(t) \;-\; a_{\alpha,\beta}(t)|\,dt = \mathbf{o}(e^x)\,. \tag{5.11}$$

Thus

$$\int |\{s_{\alpha,\beta+1}(t) \;-\; s_{\alpha,\beta}(t)\} - \{s_{\alpha,\beta+\alpha+1}(t) \;-\; s_{\alpha,\beta+\alpha}(t)\}|\,dt = \mathbf{o}(e^x)\,,$$

and further, in view of lemma 5.11, we have that

$$\int |\{s_{\alpha,\beta+1}(t) \;-\; s_{\alpha,\beta}(t)\} - \{s_{\alpha,\beta+\delta+1}(t) \;-\; s_{\alpha,\beta+\delta}(t)\}|\,dt = \mathbf{o}(e^x)\,,$$

where, as in theorem 5.23, $\delta = k\alpha$.

Also starting with (5.11), and arguing as in the second part of the proof of theorem 5.24, we obtain that

$$\int_0^x |s_{\alpha,\beta+\delta+1}(t) \;-\; s_{\alpha,\delta}(t)|\,dt = \mathbf{o}(e^x)\,.$$

Hence

$$\alpha\int_0^x |s_{\alpha,\beta+\delta+1}(t) - s_{\alpha,\delta}(t)|\,dt = \int_0^x e^t|S'_{\alpha,\beta}(t)|\,dt$$
$$= \mathbf{o}(e^x)\,,$$

so that $s_n \to \sigma\,[B,\alpha,\beta+1]$.

If we now assume that $s_n \to \sigma\,[B,\alpha,\beta+1]$, we then obtain, from theorem 5.24, that $s_n \to \sigma\,[B',\alpha,\beta+1]$ and $a_n \to 0\,[B',\alpha,\beta+1]$. Thus we have both

$$\int_0^x e^t|T'_{\alpha,\beta+1}(t)|\,dt = \alpha\int_0^x |a_{\alpha,\beta+1}(t) - a_{\alpha,\beta}(t)|\,dt = \mathbf{o}(e^x)\,,$$

and

$$\int_0^x e^t|A'_{\alpha,\beta+1}(t)|\,dt = \alpha\int_0^x |a_{\alpha,\beta+1}(t)|\,dt = \mathbf{o}(e^x)\,.$$

It now follows immediately that

$$\int_0^x e^t|A'_{\alpha,\beta}(t)|\,dt = \alpha\int_0^x |a_{\alpha,\beta}(t)|\,dt = \mathbf{o}(e^x)\,,$$

and the proof is complete.

Note. It is interesting to observe that the proof of theorem 5.27 nowhere uses the full strength of the hypothesis $s_n \to \sigma\,[B',\alpha,\beta]$. In fact the weaker hypothesis $s_n \to \sigma\,(B',\alpha,\beta)$ is sufficient, for by theorem 5.24 and lemma 5.11, we can obtain that

$$\int_0^x |s_{\alpha,\beta}(t) - s_{\alpha,\beta-1}(t) - s_{\alpha,\beta+\delta}(t) + s_{\alpha,\beta+\delta-1}(t)|\,dt = \mathbf{o}(e^x)\,.$$

This gives rise to the following interesting result:

Theorem 5.28. (see Borwein and Shawyer [50], theorem 19).
$s_n \to \sigma\,[B,\alpha,\beta]$ *if and only if* $s_n \to \sigma\,(B',\alpha,\beta)$ *and* $a_n \to 0\,[B,\alpha,\beta]$.

We now state the last theorem in this section.

Theorem 5.29. (see Borwein and Shawyer [50], theorem 17).
$s_n \to \sigma\,|B',\alpha,\beta|$ *if and only if* $s_n \to \sigma\,|B,\alpha,\beta+1|$.

Proof. The ordinary summability part of this theorem is given by theorem 5.26. It remains to consider the bounded variation conditions.

Suppose first that $A_{\alpha,\beta}(x) \in BV_x[0,\infty)$. Using the argument in theorem 5.25, we obtain that

$$e^{-x} s_{\alpha,\beta+\delta}(x) \in BV_x[0,\infty)\,,$$

where $\delta = k\alpha$ as before. Thus, in view of lemma 5.10 (iii), it follows that

$$e^{-x} s_{\alpha,\beta+\delta+1}(x) \in BV_x[0,\infty)\,.$$

Using

$$\int_0^x e^{-t} a_{\alpha,\beta}(t)\,dt \in BV_x[0,\infty)$$

and lemma 5.10 (iv), we now obtain that

$$\int_0^x e^{-t} a_{\alpha,\beta+1}(t)\,dt \in BV_x[0,\infty)\,.$$

It now follows from (5.9) that

$$e^{-x} a_{\alpha,\beta+1}(x) \in BV_x[0,\infty)\,.$$

This gives that

$$e^{-x}\left\{s_{\alpha,\beta+1}(x) - s_{\alpha,\beta+\alpha+1}(x)\right\} \in BV_x[0,\infty)\,,$$

and further, since $\alpha > 0$ and using lemma 5.10 (iii), that

$$e^{-x}\left\{s_{\alpha,\beta+1}(x) - s_{\alpha,\beta+\delta+1}(x)\right\} \in BV_x[0,\infty)\,,$$

yielding that

$$\int_0^x e^{-t} s_{\alpha,\beta+1}(t)\,dt \in BV_x[0,\infty)\,.$$

This completes the first part of the proof.

The reverse direction is much simpler. We have both

$$e^{-x} a_{\alpha,\beta+1}(x) \in BV_x[0,\infty)$$

and

$$\int_0^x e^{-t} s_{\alpha,\beta+1}(t)\,dt \in BV_x[0,\infty)\,.$$

Using (5.9), we immediately obtain that

$$\int_0^x e^{-t} s_{\alpha,\beta}(t)\, dt \in BV_x[0,\infty)\,.$$

The proof is now complete.

5.6 Tauberian relationships with respect to the parameter β

In the previous sections, we have considered increases in the parameter β. Theorems such as 5.18 are essentially of Abelian type in that they require no extra conditions. The theorems presented in this section consider decreases in the parameter β. As such, they are Tauberian in nature. For a fuller discussion of Tauberian theorems, see chapters 7 and 8.

We need to make use of two new ideas here: one is that of Borel-type boundedness and the other, that of slowly decreasing.

Definition 5.30. Borel-type boundedness (see Borwein and Shawyer [50])

(i) *If* $S_{\alpha,\beta}(x) = \mathbf{O}(1)$,
then we say that $s_n = \mathbf{O}(1)\,(B,\alpha,\beta)$.

(ii) *If* $A_{\alpha,\beta}(x) = \mathbf{O}(1)$,
then we say that $s_n = \mathbf{O}(1)\,(B',\alpha,\beta)$.

Definition 5.31. *Suppose that* $\phi(x)$ *is real valued, continuous, and unboundedly increasing. A real valued function* $f(x)$ *defined on* $[0,\infty)$ *is said to be* slowly decreasing *with respect to* $\phi(x)$ *if for every* $\epsilon > 0$*, there exist positive numbers* X, δ *such that* $f(x) - f(y) > -\epsilon$ *whenever* $x \ge y \ge X$ *and* $0 < \phi(x) - \phi(y) \le \delta$.

Alternatively, this is equivalent to:

$$\liminf\,(f(x) - f(y)) \ge 0 \quad \textit{as} \quad x > y \to \infty \quad \textit{and} \quad \phi(x) - \phi(y) \to 0\,.$$

We shall prove the following results:

Theorem 5.32. (see Borwein and Smet [52], theorem 1).
If $s_n \to \sigma\,(B,\alpha,\beta+\epsilon)$ *for some* $\epsilon > 0$ *and* $a_n \to 0\,(B,\alpha,\beta)$*, then* $s_n \to \sigma\,(B,\alpha,\beta)$.

Theorem 5.33. (see Borwein and Smet [52], theorems 2 and 2*). *If* $s_n \to \sigma\,(B,\alpha,\beta+\epsilon)$ *for some* $\epsilon > 0$ *and either* $s_n = \mathbf{O}(1)\,(B,\alpha,\beta)$ *or* $a_n = \mathbf{O}(1)\,(B,\alpha,\beta)$*, then* $s_n \to \sigma\,(B,\alpha,\beta+\delta)$. *for any* $\delta > 0$.

Theorem 5.34. (see Borwein and Smet [52], theorems 3 and 3*). *If $s_n \to \sigma\,(B,\alpha,\beta+\epsilon)$ for some $\epsilon > 0$ and either $S_{\alpha,\beta}(x)$ or $A_{\alpha,\beta}(x)$ is slowly decreasing with respect to x, then $s_n \to \sigma\,(B,\alpha,\beta)$.*

Theorem 5.35. (see Borwein and Smet [52], theorem 4). *If $s_n = \mathbf{O}(1)\,(B,\alpha,\beta+\epsilon)$ for some $\epsilon > 0$ and $s_n > -K$ for all $n \geq 0$ and where K is a positive constant, then $s_n = \mathbf{O}(1)\,(B,\alpha,\beta)$.*

Theorem 5.36. (see Borwein and Smet [52], theorems 5 and 5*). *If $s_n \to \sigma\,(B,\alpha,\beta+\epsilon)$ for some $\epsilon > 0$ and either $s_n > -K$ for all $n \geq 0$ and where K is a positive constant or $a_n > -K$ for all $n \geq 0$ and where K is a positive constant, then $s_n \to \sigma\,(B,\alpha,\beta)$.*

5.6.1 Preliminary results

Lemma 5.37. (see Borwein and Smet [52], lemmas 1 and 2)

(i) $$S_{\alpha,\beta+\delta}(x) = \int_0^x h(x-t)S_{\alpha,\beta}(t)\,dt$$

where $\delta > 0$ *and* $$h(u) = \frac{u^{\delta-1}e^{-u}}{\Gamma(\delta)},$$

(ii) $A_{\alpha,\beta}(x) = S_{\alpha,\beta}(x) - S_{\alpha,\beta+\delta}(x) + \mathbf{o}(1)$ *as* $x \to \infty$.

(iii) *If* $\displaystyle\sum_{n=0}^{\infty} a_n = \sigma\,(B,\alpha,\beta)$ *then* $a_n \to 0\,(B,\alpha,\beta)$.

The proofs are straightforward.

Lemma 5.38. (see Borwein and Smet [52], theorem 8). *Let $f(t)$ be Lebesgue integrable on every finite subinterval of $[0,\infty)$, and let $F(x) = \displaystyle\int_0^x e^{-(x-t)} f(t)\,dt$.*

If $F(x) \to \sigma$ as $x \to \infty$ and $f(t)$ is slowly decreasing, then $f(x) \to \sigma$ as $x \to \infty$.

Proof. Define

$$g(u) = \begin{cases} f(\log(u)) & u \geq 1, \\ 0 & 0 \leq u < 1. \end{cases}$$

If $w = e^x$, then we have

$$F(x) = \frac{1}{w} \int_0^w g(u)\, du\,.$$

The result now follows from the continuous version of theorem 68 in Hardy **T**[17]. See also Widder **T**[31]. This version states that if $g(x)$ is slowly decreasing with respect to $\log(x)$, and if $g(x) \to \sigma\ (C, 1)$ then $g(x) \to \sigma$.

Lemma 5.39. (see Borwein and Smet [52], lemma 3). *Suppose that $f(t)$ is bounded on every finite subinterval of $[0, \infty)$ and is slowly decreasing with respect to x. Then there exist positive numbers M_1 and M_2 such that*

$$f(x) - f(y) \geq -M_1(x-y) - M_2$$

whenever $x \geq y \geq 0$.

Proof. Since $f(t)$ is slowly decreasing with respect to x, there exist positive numbers X and δ such that $f(x) - f(y) > -1$ if $x \geq y \geq X$ and $x - y \leq \delta$. Hence, if $x \geq y \geq X$ and m is the smallest positive integer such that $(x-y)/m \leq \delta$, then

$$\begin{aligned} f(x) - f(y) &= \sum_{j=1}^{m} \left\{ f\left(y + j\frac{x-y}{m} \right) - f\left(y + (j-1)\frac{x-y}{m} \right) \right\} \\ &> -m = -(m-1) - 1 \geq -\frac{x-y}{\delta} - 1\,. \end{aligned}$$

Thus, if $M = \sup\limits_{0 \leq x \leq X} |f(x)|$, then

$$f(x) - f(y) > -\frac{1}{\delta}(x-y) - (2M+1)$$

whenever $x \geq y \geq 0$.

There is a more general result due to Borwein and Watson [55] which we state, but do not prove here.

Lemma 5.40. *Suppose that $\phi(x)$ is a real-valued, nonnegative, increasing function defined on $[0, \infty)$ such that $\phi(x) \to \infty$ as $x \to \infty$, that $f(t)$ is bounded on every finite subinterval of $[0, \infty)$ and is slowly decreasing with respect to $\phi(x)$. Then there exist positive numbers M_1 and M_2 such that*

$$f(x) - f(y) \geq -M_1\left(\phi(x) - \phi(y)\right) - M_2$$

whenever $x \geq y \geq 0$.

Lemma 5.41. (see Borwein and Smet [52], theorem 9). *Suppose that $h(u)$ is a real-valued nonnegative Lebesgue integrable function such that*

$$0 < \int_0^\infty h(u)\,du < \infty \quad \textit{and} \quad \int_0^\infty uh(u)\,du < \infty .$$

Suppose that $f(t)$ is a real-valued function such that, for some positive numbers M_1 and M_2,

$$f(x) - f(y) \geq -M_1(x-y) - M_2$$

whenever $x \geq y \geq 0$, and such that, for all $x \geq 0$,

$$F(x) = \int_0^x h(x-t) f(t)\,dt$$

exists as a Lebesgue integral. If $F(x)$ is bounded on $[0, \infty)$, then $f(x)$ is bounded on $[0, \infty)$.

Proof. Suppose that $M_3 = \sup_{x \geq 0} |F(x)| < \infty$. Choose X such that

$$L = \int_0^X h(u)\,du > 0 .$$

Now

$$\begin{aligned} f(x) \int_0^x h(x-t)\,dt &= \int_0^x h(x-t)\{f(x) - f(t)\}\,dt + F(x) \\ &\geq \int_0^x h(x-t)\{-M_1(x-t) - M_2\}\,dt + F(x) \\ &\geq -M_1 \int_0^\infty uh(u)\,du - \int_0^\infty h(u)\,du - M_3 \\ &= -M_4 \quad \text{say.} \end{aligned}$$

It follows that $f(x) \geq -M_4/L$ if $x \geq X$. But $f(x) \geq -M_1 X - M_2 + f(0)$ for $0 \leq x \leq X$. Hence there exists a positive number M_5 such that $f(x) \geq -M_5$ for all $x \geq 0$.

If $x \geq X$, then we have

$$M_3 \geq F(x) = \int_0^{x-X} h(x-t)f(t)\,dt + \int_{x-X}^{x} h(x-t)f(t)\,dt$$

$$\geq -M_5 \int_o^{x-X} h(x-t)\,dt$$

$$+ \int_{x-X}^{x} h(x-t)\{f(x-X) - M_1(t-x+X) - M_2\}\,dt$$

$$\geq M_6 + f(x-X)\int_0^X h(u)\,dy$$

where

$$M_6 = -M_5 \int_X^\infty h(u)\,du + M_1 \int_0^X uh(u)\,du - (M_1X + M_2)\int_0^X h(u)\,du\,.$$

It now follows that $(M_3 - M_6)/L \geq f(x)$ for $x \geq 0$, and so that $f(x)$ is bounded on $[0,\infty)$.

Lemma 5.42. (see Borwein and Smet [52], theorem 10). *Suppose that $f(t)$ is a real-valued non-decreasing function defined on $[0,\infty)$, that $\delta > 0$, and that $e^{-x}f_\delta(x)$ is bounded on $[0,\infty)$. Then $e^{-x}f(x)$ is also bounded on $[0,\infty)$.*

Proof. Write $F(x) = e^{-x}f_\delta(x)$ and $M = \sup_{x\geq 0}|F(x)| < \infty$. Since $f(t)$ is nondecreasing, we have, for all $x \geq 0$,

$$\delta\,\Gamma(\delta)\,e\,M \geq \delta\,\Gamma(\delta)\,e\,F(x+1)$$

$$= \delta\,e^{-x}\int_0^x (x+1-t)^{\delta-1}f(t)\,dt$$

$$+\delta\,e^{-x}\int_x^{x+1} (x+1-t)^{\delta-1}f(t)\,dt$$

$$\geq \delta\,e^{-x}\int_0^x (x+1-t)^{\delta-1}f(0)\,dt$$

$$+\delta\,e^{-x}\int_x^{x+1} (x+1-t)^{\delta-1}f(x)\,dt$$

$$= f(0)\left\{(x+1)^\delta - 1\right\} e^{-x} + e^{-x} f(x)\,,$$

so that it follows that $e^{-x}f(x)$ is bounded on $[0,\infty)$.

5.6.2 Proofs of the theorems

Proof of theorem 5.32. Let k be a positive integer. Then, since $a_n \to 0\,(B,\alpha,\beta)$, we have, by theorem 5.18, that $A_{\alpha,\beta+(k-1)\alpha}(x) \to 0$ as $x \to \infty$. Further, using lemma 5.37 (ii),

$$S_{\alpha,\beta+(k-1)\alpha}(x) = A_{\alpha,\beta+(k-1)\alpha}(x) \,+\, S_{\alpha,\beta+k\alpha}(x) \,+\, \mathbf{o}(1)$$

as $x \to \infty$.

Since $s_n \to \sigma\,(B,\alpha,\beta+\epsilon)$, we may choose an integer k such that $s_n \to \sigma\,(B,\alpha,\beta+k\alpha)$, so that $S_{\alpha,\beta+k\alpha} \to \sigma$ as $x \to \infty$. It now follows that $S_{\alpha,\beta+(k-1)\alpha} \to \sigma$ as $x \to \infty$. Repeating this step results in $S_{\alpha,\beta} \to \sigma$ as $x \to \infty$ and the proof is complete.

Proof of theorem 5.33. First suppose that $s_n = \mathbf{O}(1)\,(B,\alpha,\beta)$. Further, we may suppose without loss of generality that $\{s_n\}$ is a real sequence. Let $\delta > 0$ and define $M = \sup\limits_{x\geq 0} |S_{\alpha,\beta}(x)|$.

Let k be a positive integer and define $h(u) = \dfrac{u^{(k-2)+\delta}e^{-u}}{\Gamma(k-1+\delta)}$. Then, by lemma 5.37 (i), we have for $x \geq y \geq 0$, that

$$\begin{aligned}
&\left|S_{\alpha,\beta+(k-1)+\delta}(x) - S_{\alpha,\beta+(k-1)+\delta}(y)\right| \\
&= \left|\int_y^x h(x-t)S_{\alpha,\beta}(t)\,dt \,+\, \int_0^y \{h(x-t)-h(y-t)\}\, S_{\alpha,\beta}(t)\,dt\right| \\
&\leq M\int_y^x h(x-t)\,dt \,+\, M\int_0^y |h(x-t)-h(y-t)|\,dt \\
&\leq M\int_0^{x-y} h(u)\,du \,+\, M\int_0^\infty |h(x-y+u)-h(u)|\,du
\end{aligned}$$

as $x - y \to 0$, since $h(u)$ is Lebesgue integrable on $[0,\infty)$. Further, by lemma 5.37 (i), we have that

$$S_{\alpha,\beta+k+\delta}(x) = \int_0^x e^{-(x-t)} S_{\alpha,\beta+(k-1)+\delta}(t)\,dt\,.$$

From theorem 5.18, we have that $S_{\alpha,\beta+k+\delta} \to \sigma$ as $x \to \infty$ whenever $k+\delta \geq \epsilon$. Further, by lemma 5.38 with $F(x) = S_{\alpha,\beta+k+\delta}(x)$ and $f(x) = S_{\alpha,\beta+(k-1)+\delta}(x)$, we have that $S_{\alpha,\beta+(k-1)+\delta} \to \sigma$ as $x \to \infty$ whenever $S_{\alpha,\beta+k+\delta} \to \sigma$ as $x \to \infty$. Repeating this step results in $S_{\alpha,\beta+\delta} \to \sigma$ as

$x \to \infty$ and the proof of this case is complete.

Now suppose that $a_n = \mathbf{O}(1)\,(B, \alpha, \beta)$. By lemma 5.37 (iii), we have that $a_n \to 0\,(B, \alpha, \beta + \epsilon)$, and further, by the first part of this proof, that $a_n \to 0\,(B, \alpha, \beta + \delta)$ for any $\delta > 0$. The conclusion follows from theorem 5.32 above.

Proof of theorem 5.34. First suppose that $S_{\alpha,\beta}(x)$ is slowly decreasing with respect to x. Using lemma 5.37 (i), lemma 5.39, and lemma 5.41 with $F(x) = S_{\alpha,\beta+\epsilon}(x)$, $f(x) = S_{\alpha,\beta}(x)$ and $h(u) = u^{\epsilon-1}e^{-u}/\Gamma(\epsilon)$, we have that $S_{\alpha,\beta}(x)$ is bounded on $[0, \infty)$. Hence, by theorem 5.33, $S_{\alpha,\beta+1}(x) \to \sigma$ as $x \to \infty$.

Using lemma 5.37 (i) and lemma 5.38 with $F(x) = S_{\alpha,\beta+1}(x)$ and $f(x) = S_{\alpha,\beta}(x)$, we have that $S_{\alpha,\beta}(x) \to \sigma$ as $x \to \infty$ and the proof of this part is complete.

If we suppose that $A_{\alpha,\beta}(x)$ is slowly decreasing with respect to x, the proof is similar.

Proof of theorem 5.35. In view of lemma 5.37 (i), we can assume without loss of generality that $\epsilon = \delta$. The result now follows from lemma 5.42 with

$$F(x) = S_{\alpha,\beta+\delta}(x) + \alpha\, e^{-x} \sum_{n=N}^{\infty} K \frac{x^{\alpha n+\beta+\delta-1}}{\Gamma(\alpha n + \beta + \delta)},$$

$$e^{-x} f(x) = S_{\alpha,\beta}(x) + \alpha\, e^{-x} \sum_{n=N}^{\infty} K \frac{x^{\alpha n+\beta-1}}{\Gamma(\alpha n + \beta)}.$$

Proof of theorem 5.36. First suppose that $s_n > -K$ for all $n \geq 0$ and where K is a positive constant. Let k be a positive integer such that $k > \epsilon$. The by theorem 5.35, we have that $s_n = \mathbf{O}(1)\,(B, \alpha, \beta + \epsilon - k)$, and further, by theorem 5.33, that $s_n \to \sigma\,(B, \alpha, \beta)$.

The proof for the case in which $a_n > -K$ for all $n \geq 0$ and where K is a positive constant is similar.

Further results of the types illustrated here can be found in the literature. In particular, see Borwein and Smet [52] and [53].

5.7 Extended definitions

Definition 5.43. Absolute summability with indices $p > 0$ and real λ (see Shawyer [248])

(i) *If* $A_{\alpha,\beta}(x) \to \sigma_N$ *as* $x \to \infty$, *and* $\displaystyle\int_0^\infty e^{tp\lambda} \left|A'_{\alpha,\beta}(x)\right|^p dt < \infty$, *then we say that* $s_n \to \sigma\,|B', \alpha, \beta|_{p,\lambda}$.

(ii) *If* $S_{\alpha,\beta}(x) \to \sigma_N$ *as* $x \to \infty$, *and* $\int_0^\infty e^{tp\lambda} \left|S'_{\alpha,\beta}(x)\right|^p dt < \infty$, *then we say that* $s_n \to \sigma\,|B, \alpha, \beta|_{p,\lambda}$.

Definition 5.44. Strong summability with index $p > 0$: (see Borwein and Shawyer [51])

(i) *If* $\int_0^x e^t \left|A_{\alpha,\beta-1}(t) - \sigma_N\right|^p dt = \mathbf{o}(e^x)$, *then we say that* $s_n \to \sigma\,[B', \alpha, \beta]_p$.

(ii) *If* $\int_0^x e^t \left|S_{\alpha,\beta-1}(t) - \sigma\right|^p dt = \mathbf{o}(e^x)$, *then we say that* $s_n \to \sigma\,[B, \alpha, \beta]_p$.

Theorems similar to those in the previous section can be found in Borwein and Shawyer [51], and in Shawyer [248] and [247].

In the proofs of these theorems, use of the triangle inequality is replaced by use of Minkowski's inequality. Much use is also made of Hölder's inequality. We shall therefore adopt the standard notation for the number p', conjugate to the number p, as given by

$$\frac{1}{p} + \frac{1}{p'} = 1\,.$$

5.7.1 Results involving strong summability

The following theorems are stated here without proof.

Theorem 5.45. (see Borwein and Shawyer [51], theorem 3*). *If* $s_n \to \sigma\,[B, \alpha, \beta]_p$ *then* $s_n \to \sigma\,(B, \alpha, \beta - \delta)$ *where* $p > 1$ *and* $\delta < (p-1)/p = 1/p'$.

Note that an appropriate use of Hölder's inequality allows the reduction in the parameter β.

Theorem 5.46. (see Borwein and Shawyer [51], theorem 5*). *If* $s_n \to \sigma\,(B, \alpha, \beta)$ *then* $s_n \to \sigma\,[B, \alpha, \beta + 1]_p$ *where* $p > 0$.

Theorem 5.47. (see Borwein and Shawyer [51], theorem 9*). *If* $s_n \to \sigma\,[B, \alpha, \beta]_p$ *then* $s_n \to \sigma\,[B, \alpha, \beta + \delta]_q$ $(\delta > 0)$ *provided that one of the following conditions it satisfied:*

(i) $p > q > 0\,,\,\delta = 0\,,$

(ii) $q \geq p \geq 1\,,\,\delta > \frac{1}{p} - \frac{1}{q}\,,$

(iii) $q > p > 1\,,\,\delta = \frac{1}{p} - \frac{1}{q}\,.$

Theorem 5.48. (see Borwein and Shawyer [51], theorem 11*). *For* $p > 1$, $s_n \to \sigma\,[B, \alpha, \beta]_p$ *if and only if* $s_n \to \sigma\,(B, \alpha, \beta)$ *and*

$$\int_0^x e^t \left|S'_{\alpha,\beta}(t)\right|^p dt = \mathbf{o}(e^x).$$

Theorem 5.49. (see Borwein and Shawyer [51], theorem 15*). *For* $p > 1$, $s_n \to \sigma\,[B, \alpha, \beta]_p$ *if and only if* $s_n \to \sigma\,[B', \alpha, \beta]_p$ *and* $a_n \to 0\,[B, \alpha, \beta]_p$.

Theorem 5.50. (see Borwein and Shawyer [51], theorem 18*). *For* $p > 1$, $s_n \to \sigma\,[B', \alpha, \beta]_p$ *if and only if* $s_n \to \sigma\,[B, \alpha, \beta + 1]_p$.

We also can obtain a 'step back' theorem.

Theorem 5.51. (see Shawyer [247], theorem 1). *For* $p > 1$, *if* $s_n \to \sigma\,[B, \alpha, \beta + 1]_p$ *and* $\int_0^x e^t |S'_{\alpha,\beta}(t)|^p\,dt = \mathbf{o}(e^x)$, *then* $s_n \to \sigma\,[B, \alpha, \beta]_p$.

As with theorem 5.28, it is possible to weaken some of the requirements in some of these results, giving rise to the following.

Theorem 5.52. (see Shawyer [247], theorem 2). *For* $p > 1$, *if* $s_n \to \sigma\,(B, \alpha, \delta)$ *for some* δ *and* $\int_0^x e^t |S'_{\alpha,\beta}(t)|^p\,dt = \mathbf{o}(e^x)$, *then* $s_n \to \sigma\,[B, \alpha, \beta]_p$.

Theorem 5.53. (see Shawyer [247], theorem 3). *For* $p > 1$, *if* $s_n \to \sigma\,(B, \alpha, \delta)$ *for some* δ *and if* $a_n \to 0\,[B, \alpha, \beta]_p$, *then* $s_n \to \sigma\,[B', \alpha, \beta]_p$.

5.7.2 Results involving absolute summability

In this section, we assume that $p > 0, q > 0$ and that when $p > 1$ or $q > 1$, the conjugate number is denoted by p' or q' respectively. λ and μ denote real numbers.

The following theorems are stated here without proof.

Theorem 5.54. (see Shawyer [248], theorem 1). *If* $s_n \to \sigma\,|B, \alpha, \beta|_{p,\lambda}$ *then* $s_n \to \sigma\,(B, \alpha, \beta)$.

Theorem 5.55. (see Shawyer [248], theorem 2). *For* $\lambda \geq 0$ *and* $p > 1$, *if* $s_n \to \sigma\,|B, \alpha, \beta|_{p,\lambda}$ *then* $s_n \to \sigma\,[B, \alpha, \beta]_p$.

Theorem 5.56. (see Shawyer [248], theorem 3). *If* $s_n \to \sigma\,|B, \alpha, \beta|_{q,\lambda}$ *then* $s_n \to \sigma\,|B, \alpha, \beta + \delta|_{p,\mu}$ *whenever*

(1)	$\delta = 0$	(a)	$p = q$ *and* $\mu \le \lambda$;
		(b)	$p < q$ *and* $\mu < \lambda$;
(2)	$\delta > \frac{1}{q}$	(a)	$q > 1$ *and* $\mu < \lambda < 1$;
	$\delta > 1 - \frac{1}{p}$	(b)	$p \ge q = 1, \mu \le \lambda$ *and* $\mu < 1$;
(3)	$\mu = \lambda < 1$	(a)	$\delta > \frac{1}{q}$ *and* $p = q \ge 1$;
		(b)	$\delta = 1$ *and* $p = q = 1$.

6
Abelian theorems

6.1 Introduction

As stated in chapter 5, certain results are known as *Abelian* theorems because they are of the nature of Abel's theorem in the limit of a power series at its radius of convergence.

Theorem 6.1. *If* $\sum_{n=0}^{\infty} a_n\, x^n$ *has radius of convergence* ρ, *and if* $\sum_{n=0}^{\infty} a_n\, \rho^n$ *is convergent, then* $\lim_{x \to \rho^-} \sum_{n=0}^{\infty} a_n\, x^n = \sum_{n=0}^{\infty} a_n\, \rho^n$.

6.2 Abelian theorems for ordinary Borel-type methods

The theorems in this section are analogues of simpler theorems to be found in Hardy **T**[17]. We will prove theorems for (B', α, β) summability, and since the results are independent of the value of β, the theorems are also true for (B, α, β) summability.

Theorem 6.2. (see Shawyer [249], theorem 1). *If the series* $\sum_{n=0}^{\infty} a_n X^n$ *is summable* (B', α, β) *at a point* P *in the complex plane, then it is summable* (B', α, β) *at every point on* OP *(the line segment joining* P *to the origin* O*).*

The case $\alpha = \beta = 1$ is due to Phragmén [216] and the case $\alpha > 0, \beta = 1$ is due to Mittag-Leffler [194]. See also theorem 127 in Hardy **T**[17].

Proof. We may assume, without loss of generality, that P is the unit point 1. Then the hypothesis of the theorem is that

$$J(X) = \int_0^\infty e^{-t}\,dt \sum_{n=N}^\infty \frac{a_n X^n t^{\alpha n+\beta-1}}{\Gamma(\alpha n+\beta)},$$

is convergent for $X = 1$, and it is required to prove it convergent for $0 < X < 1$. For convenience, make the transformation $X = x^\alpha$. Thus

$$\begin{aligned} J(X) = I(x) &= \int_0^\infty e^{-t}\,dt \sum_{n=N}^\infty \frac{a_n x^{\alpha n} t^{\alpha n+\beta-1}}{\Gamma(\alpha n+\beta)} \\ &= x^{1-\beta}\int_0^\infty e^{-t} a_{\alpha,\beta}(xt)\,dt \\ &= x^{-\beta}\int_0^\infty e^{(1-1/x)u} e^{-u} a_{\alpha,\beta}(u)\,du\,. \end{aligned}$$

The hypothesis is now that $I(x)$ is convergent for $x = 1$. It is easy to see that $I(x)$ is convergent for $0 < x < 1$.

In fact, following Phragmén [216], it is easy to see that $I(x)$ is uniformly convergent for $0 < \delta \leq x \leq 1$. But this is not the full picture. The next theorem shows that the convergence is in fact uniform for $0 \leq x \leq 1$.

Theorem 6.3. (see Shawyer [249], theorem 2). *If the series* $\sum_{n=0}^\infty a_n X^n$ *is summable* (B', α, β) *at a point* P *in the complex plane, then it is uniformly summable* (B', α, β) *on* OP.

The case $\alpha = \beta = 1$ is due to Hardy [108]. See also theorem 130 in Hardy **T**[17].

Proof. As in the proof of theorem 6.2, we may assume that P is the unit point 1. Further, for convenience in the calculations, we assume that all the a_n are real. There are two cases to consider: case (i) $\beta \leq 0$, and case (ii) $\beta > 0$.

Case (i): $\beta \leq 0$. It is sufficient to show that, given any $\epsilon > 0$, there corresponds a number $H_0(\epsilon) > 0$ such that

$$|I_0(x)| = \left| x^{1-\beta}\int_H^G e^{-t} a_{\alpha,\beta}(xt)\,dt \right| < \epsilon\,,$$

whenever $G > H > H_0(\epsilon)$, and $0 \leq x \leq 1$. Thus we have

$$\begin{aligned}|I_0(x)| &= \left| x^{-\beta} \int_{Hx}^{Gx} e^{(1-1/x)u} e^{-u} a_{\alpha,\beta}(u)\, du \right| \\ &\le x^{-\beta} e^{(1-1/x)Hx} \sup_{y>Hx} \left| \int_{Hx}^{y} e^{-u} a_{\alpha,\beta}(u)\, du \right| ,\end{aligned}$$

and so, for $0 \le x \le \frac{1}{2}$,

$$\begin{aligned}|I_0(x)| &\le x^{-\beta} e^{-(1-x)H} \sup_{y>z>0} \left| \int_{z}^{y} e^{-u} a_{\alpha,\beta}(u)\, du \right| \\ &\le M e^{-H/2} < \epsilon\end{aligned}$$

for $H > H_0(\epsilon)$, and, for $\frac{1}{2} \le x \le 1$,

$$\begin{aligned}|I_0(x)| &\le x^{-\beta} e^{-(1-x)H} \sup_{y>z>H/z} \left| \int_{z}^{y} e^{-u} a_{\alpha,\beta}(u)\, du \right| \\ &< \epsilon\end{aligned}$$

for $H > H_0(\epsilon)$.
Case (ii): $\beta > 0$. Define

$$\delta = \begin{cases} \dfrac{1}{2\beta} & \text{whenever } \beta \ge 1\,, \\ \dfrac{1}{2} & \text{whenever } 0 < \beta < 1\,. \end{cases}$$

There are three subcases to consider:

$$\text{(a) } Gx \le 1\,; \qquad \text{(b) } Hx < 1 < Gx\,; \qquad \text{(c) } Hx \ge 1\,.$$

Assume without loss of generality that $H > 1/\delta \ge 2\beta$ and define

$$M = \max_{y>1} \left| \int_1^y e^{-t} a_{\alpha,\beta}(t)\, dt \right| .$$

We can write $I_0(x) = I_1 + I_2$, where

$$I_1 = x^{1-\beta} \int_H^{1/x} e^{-t} a_{\alpha,\beta}(xt)\, dt\,,$$

and

$$I_2 = x^{1-\beta} \int_{1/x}^{G} e^{-t} a_{\alpha,\beta}(xt)\, dt\,.$$

Subcase (b): First we note that it is easy to show that if the k_n are real numbers for $n = N, N+1, \ldots$ and if $g(t) = \sum_{n=N}^{\infty} k_n t^n$ is convergent for $0 \leq t \leq T$, then $g(t) = \mathbf{O}(t^N)$ for $0 \leq t \leq T$.

Thus, since $a_{\alpha,\beta}(u)$ is a convergent series for all $u \geq 0$, it follows that $|a_{\alpha,\beta}(xt)| < M_0\,(xt)^{\alpha N+\beta-1}$ for $xt \leq 1$, where M_0 is a constant. Therefore we have

$$\begin{aligned} |I_1| &< M_0\, x^{\alpha N} \int_{H}^{1/x} e^{-t} t^{\alpha N+\beta-1}\, dt \\ &< M_0\, \Gamma(\alpha N+\beta)\, x^{\alpha N} \\ &< M_0\, \Gamma(\alpha N+\beta)\, H^{-\alpha N}\,. \end{aligned} \tag{6.1}$$

Further

$$\begin{aligned} I_1 &= x^{1-\beta} \int_{1/x}^{G} e^{-t} a_{\alpha,\beta}(xt)\, dt \\ &= x^{-\beta} \int_{1}^{Gx} e^{u(1-1/x)} e^{-u} a_{\alpha,\beta}(u)\, du \\ &= x^{-\beta} e^{(1-1/x)} \int_{1}^{y} e^{-u} a_{\alpha,\beta}(u)\, du\,, \end{aligned}$$

where $1 < y < Gx$, and so

$$|I_2| \leq M\, x^{-\beta} e^{(1-1/x)} \leq M\, x^{-\beta} e^{-\frac{1}{2}x} < M\, H^{\beta} e^{-H/2}\,, \tag{6.2}$$

since $x \leq \delta \leq \frac{1}{2}, 1/x \geq H > 2\beta$ and $w^{\beta} e - w/2$ is monotonic decreasing for $w > 2\beta$. Thus, since the right-hand sides of (6.1) and (6.2) both tend to zero as $H \to \infty$, it follows at once that $|I_0(x)| < \epsilon$ whenever $H > H_0(\epsilon)$. This completes the proof for subcase (b).

Subcases (a) and (c): For subcase (a), we note that the range of $I_0(x)$ is contained in that of I_1 of subcase (b), and so can be treated as above. Similarly, for subcase (c), the range of $I_0(x)$ is contained in that of I_2 of subcase (b), and so also can be treated as above.

This completes the proof of theorem 6.3.

Theorem 6.4. (see Shawyer [249], theorem 3). *If the series* $\sum_{n=0}^{\infty} a_n X^n$ *is summable* (B', α, β) *at a point* P *in the complex plane, then the* (B', α, β) *sum of the series on* OP *is an analytic function of* x, *regular inside the circle in the* x*-plane described on* $\wr_{\surd}$ *as diameter, where* $\wr_{\surd}$ *is the image of* OP.

Note that $x = X^{1/\alpha}$ has its principal value on OP.

This theorem can be deduced from theorem 6.3 above by exactly the same method as is used to deduce theorem 132 from 131 in Hardy **T**[17], and as is used to deduce theorem 2 from theorem 1 in Mittag-Leffler [194].

This gives rise to Borel-type polygons, discs of regularity and diameters of summability in the new variable x; when transformed back to the original variable X, the *polygon* will become a *curvilinear polygon* depending on α but not on β. For a descriptive analysis, see Hardy **T**[17], pp. 197–198, or Mittag-Leffler [194]. Also, Borel's monograph **T**[2] or **T**[3] contains a chapter discussing polygons of summability.

6.3 Abelian theorems for strong Borel-type methods

The theorem here is proved for $[B', \alpha, \beta]_p$ summability, and since the results are independent of the values of β, it is also true for $[B, \alpha, \beta]_p$ summability.

Theorem 6.5. (see Shawyer [249], theorem 1S). *Suppose that* $q \geq 1$. *If the series* $\sum_{n=0}^{\infty} a_n X^n$ *is summable* $[B', \alpha, \beta]_q$ *at a point* P *in the complex plane, then it is summable* $[B', \alpha, \beta]_q$ *at every point on* OP.

Proof. Again, suppose that P is the unit point 1. The ordinary summability part is given by theorem 6.2 above. It is therefore sufficient to prove that, for $0 < x < 1$,

$$\int_0^y e^t \left| \frac{d}{dt} \left(x^{1-\beta} \int_0^t e^{-u} a_{\alpha,\beta}(xu)\, du \right) \right|^q dt = \mathbf{o}(e^y)$$

as $y \to \infty$, whenever

$$\int_0^y e^t \left| A'_{\alpha,\beta}(t) \right|^q dt = \mathbf{o}(e^y)$$

as $y \to \infty$.

Alternatively, it is sufficient to prove that

$$x^{(1-\beta)q}\int_0^y |a_{\alpha,\beta}(xt)|^q\,dt = \mathbf{o}(e^{qy})$$

as $y \to \infty$, whenever

$$\int_0^y |a_{\alpha,\beta}(t)|^q\,dt = \mathbf{o}(e^{qy})$$

as $y \to \infty$. Now

$$\begin{aligned} e^{-qy}x^{(1-\beta)q}\int_0^y |a_{\alpha,\beta}(xt)|^q\,dt &= e^{-qy}x^{(1-\beta)q-1}\int_0^{xy} |a_{\alpha,\beta}(u)|^q\,du \\ &= e^{-qy}x^{(1-\beta)q-1}\mathbf{o}(e^{xyq}) \\ &= \mathbf{o}(1) \end{aligned}$$

as $y \to \infty$ for each fixed $x \in (0,1)$. The proof is now complete.

6.4 Abelian theorems for absolute Borel-type methods

The first two theorems here are proved for $|B',\alpha,\beta|$ summability, and since the results are independent of the values of β, they are also true for $|B,\alpha,\beta|$ summability.

Theorem 6.6. (see Shawyer [249], theorem 1A). *If the series* $\sum_{n=0}^{\infty} a_n X^n$ *is summable* $|B',\alpha,\beta|$ *at a point* P *in the complex plane, then it is summable* $|B',\alpha,\beta|$ *at every point on* OP.

Proof. Again, suppose that P is the unit point 1. The ordinary summability part is given by theorem 6.2 above. The extra hypothesis, when P is the point 1, is that

$$\int_0^\infty e^{-t}|a_{\alpha,\beta}(t)|\,dt < \infty,$$

and it is required to prove that, for each $x \in (0,1)$,

$$K_{\alpha,\beta}(x) := x^{1-\beta}\int_0^\infty e^{-t}|a_{\alpha,\beta}(xt)|\,dt < \infty\,.$$

The proof can now be completed by replacing '$a_{\alpha,\beta}(.)$' with '$|a_{\alpha,\beta}(.)|$' in the proof of theorem 6.2 above.

Theorem 6.7. (see Shawyer [249], theorem 2A). *If the series* $\sum_{n=0}^{\infty} a_n X^n$ *is summable* $|B', \alpha, \beta|$ *at a point* P *in the complex plane, then* $K_{\alpha,\beta}(X)$ *is a convergent integral, uniformly in* X *on* OP.

Proof. The convergence of $K(X)$ on OP follows from theorem 6.6 above. The proof can now be completed by replacing '$a_{\alpha,\beta}(.)$' with '$|a_{\alpha,\beta}(.)|$' in the proof of theorem 6.3 above.

We conclude this section with a theorem of a different type.

Theorem 6.8. (see Shawyer [249], theorem 4). *If the series* $\sum_{n=0}^{\infty} a_n X^n$ *is summable* $|B', \alpha, \beta|$ *at a point* P *in the complex plane, then*

$$J(X) := X^{1-\beta} \int_0^\infty e^{-t} a_{\alpha,\beta}(Xt)\, dt\,,$$

is of bounded variation with respect to X *on* OP.

Proof. Assume that P is the unit point 1 so that we have

$$\int_0^\infty e^{-t} |a_{\alpha,\beta}(t)|\, dt < \infty\,.$$

We also note that, since $\alpha N > 1$, $t^{-\beta} a_{\alpha,\beta}(t)$ is continuous for all $t \geq 0$. Further, since $I(x)$ is continuous on $[0,1]$ (by theorem 6.7 above) and has a continuous derivative on $(0,1)$, it is sufficient, for the proof of this theorem, to show that

$$\int_0^1 |I'(x)|\, dx < \infty\,.$$

Now

$$I(x) = x^{-\beta} \int_0^\infty e^{-u/x} a_{\alpha,\beta}(u)\, du\,,$$

and so

$$\begin{aligned} I'(x) &= x^{-1-\beta} \int_0^\infty e^{-u/x} \left(\frac{u}{x} - \beta \right) a_{\alpha,\beta}(u)\, du \\ &= x^{-\beta} \int_0^\infty e^{-t} (t - \beta) a_{\alpha,\beta}(xt)\, dt\,. \end{aligned}$$

Thus

$$\int_0^1 |I'(x)|\, dx = \int_0^1 x^{-\beta}\, dx \int_0^\infty |t - \beta| e^{-t} |a_{\alpha,\beta}(xt)|\, dt$$

$$= \int_0^\infty |t-\beta| e^{-t}\, dt \int_0^1 x^{-\beta} |a_{\alpha,\beta}(xt)|\, dx$$

$$= \int_0^\infty t^{\beta-1}|t-\beta| e^{-t}\, dt \int_0^t u^{-\beta} |a_{\alpha,\beta}(u)|\, du\,.$$

There are two cases to consider:
Case (i): $\beta \le 0$. Here then we have $|t-\beta| = t-\beta$ since $t \ge 0$. Thus

$$\int_0^1 |I'(x)|\, dx \le \int_0^\infty -\frac{d}{dt}\left(t^\beta e^{-t}\right)\, dt \int_0^t u^{-\beta}|a_{\alpha,\beta}(u)|\, du$$

$$= \int_0^\infty u^{-\beta}|a_{\alpha,\beta}(u)|\, du \int_u^\infty -\frac{d}{dt}\left(t^\beta e^{-t}\right)\, dt$$

$$= \int_0^\infty e^{-u}|a_{\alpha,\beta}(u)|\, du < \infty\,.$$

Case (ii): $\beta > 0$. Write

$$\int_0^1 |I'(x)|\, dx \le I_1 + I_2\,,$$

where

$$I_1 := \int_0^\beta t^{\beta-1}(\beta - t)e^{-t}\, dt \int_u^{-\beta} |a_{\alpha,\beta}(u)|\, du$$

$$\le \int_0^\beta \frac{d}{dt}\left(t^\beta e^{-t}\right)\, dt \int_0^\beta u^{-\beta}|a_{\alpha,\beta}(u)|\, du \;<\; \infty$$

and

$$I_2 := \int_\beta^\infty t^{\beta-1}(\beta - t)e^{-t}\, dt \int_u^{-\beta} |a_{\alpha,\beta}(u)|\, du$$

$$= \int_0^\beta u^{-\beta}|a_{\alpha,\beta}(u)|\, du \int_\beta^\infty -\frac{d}{dt}\left(t^\beta e^{-t}\right)\, dt$$

$$+ \int_\beta^\infty u^{-\beta}|a_{\alpha,\beta}(u)|\, du \int_u^\infty -\frac{d}{dt}\left(t^\beta e^{-t}\right)\, dt$$

$$< \infty\,.$$

The proof is now complete.

7

Tauberian theorems – I

We begin by restating some results for Borel's methods that have been stated in chapter 3. As we develop the results in this chapter, it will become clear how these results are the origins of the more general theorems. These theorems are of three types.

(1) **The 'o' theorems**

Theorem 3.18. (see, for example, Hardy **T**[17], theorem 143). *If* $\sum_{n=0}^{\infty} a_n = \sigma\,(B)$ *and* $a_n = \mathbf{o}(n^{-\frac{1}{2}})$, *then* $\sum_{n=0}^{\infty} a_n = \sigma$.

Theorem 3.19. (see, for example, Hardy **T**[17], theorem 147). *Suppose that* $\rho \geq -\frac{1}{2}$. *If* $\sum_{n=0}^{\infty} a_n = \sigma\,(B)$ *and* $a_n = \mathbf{o}(n^{\rho})$, *then* $\sum_{n=0}^{\infty} a_n = \sigma\,(C, 2\rho{+}1)$.

(2) **The 'O' theorem.**

Theorem 3.20. (see, for example Hardy **T**[17], theorem 156). *If* $\sum_{n=0}^{\infty} a_n = \sigma\,(B)$ *and* $a_n = \mathbf{O}(n^{-\frac{1}{2}})$, *then* $\sum_{n=0}^{\infty} a_n = \sigma$.

(3) **The slowly decreasing theorem.**

Theorem 3.21. (see, for example, Hardy **T**[17], theorem 241 or **T**[22], theorems 5 and 6). *If*

(i) $\liminf\,(s_n - s_m) \geq 0$ *as* $m \to \infty$, $n > m$, $\dfrac{n-m}{\sqrt{m}} \to 0$ *and*

(ii) $\sum_{n=0}^{\infty} a_n = \sigma\,(B)$,

then $s_n \to \sigma$ *as* $n \to \infty$.

In the sections that follow, we shall give similar results for the Borel-type methods (B, α, β). Because of theorem 5.26, the results will also be true for the Borel-type methods (B', α, β).

7.1 The 'o' theorem

In this section, we prove a generalization of theorem 3.19 below. Before proceeding to the actual proof of theorem 7.1, we shall require some preliminary lemmas and some results on Cesàro sums.

Theorem 7.1. (see Borwein [43]). *If*

(i) $\rho \geq -\frac{1}{2}$, $a_n = \mathbf{o}(n^\rho)$, *and*

(ii) $\sum_{n=0}^{\infty} a_n = \sigma\,(B, \alpha, \beta)$,

then $\sum_{n=0}^{\infty} a_n$ *is summable* $(C, 2\rho+1)$ *to* σ.

The case $\alpha = \beta = 1$ of this theorem is known and is stated in chapter 3 above (theorem 3.19). Using this, and theorem 5.13 in chapter 5 above, we can immediately obtain a proof of theorem 7.1, valid for $\alpha \geq 1$. However, Borwein 's proof, given here, makes no appeal to either α or to β, and so is valid for any given α and β.

7.1.1 Preliminary results

Lemma 7.2. (see Borwein [43]).

(i) $x^\nu \Gamma(y-\nu) \geq \Gamma(y)$ *if* $x \geq y > \nu \geq 0$,

(ii) $x^\nu \Gamma(y-\nu) \leq \Gamma(y)$ *if* $\nu \geq 0,\ 0 < x \leq y - \nu - 1$.

Proof. Let $\psi(\nu) = x^\nu \Gamma(y-\nu)$. For part (i), using standard results to be found, for example, in Whittaker and Watson **T**[30], §§12.3, 12.31, we have

$$\begin{aligned}
\frac{\psi'(\nu)}{\psi(\nu)} &= \log(x) - \frac{\Gamma'(y-\nu)}{\Gamma(y-\nu)} \\
&= \log(x) - \int_0^\infty \left[\frac{e^{-t}}{t} - \frac{e^{-(y-\nu)t}}{1-e^{-t}}\right] dt \\
&\geq \log(x) - \int_0^\infty \frac{e^{-t} - e^{-(y-\nu)t}}{t}\, dt \\
&= \log(x) - \log(y-\nu) \geq 0,
\end{aligned}$$

so that $\psi(\nu) \geq \psi(0)$, as required.

Similarly, for part (ii), we have

$$\begin{aligned}\frac{\psi'(\nu)}{\psi(\nu)} &= \log(x) - \int_0^\infty \left[\frac{e^{-t}}{t} - \frac{e^{-(y-\nu)t}}{1-e^{-t}}\right] dt \\ &\leq \log(x) - \int_0^\infty \frac{e^{-t} - e^{-(y-\nu-1)t}}{t} dt \\ &= \log(x) - \log(y-\nu-1) \leq 0,\end{aligned}$$

so that $\psi(\nu) \leq \psi(0)$, as required.

Lemma 7.3. (compare with theorem 137 in Hardy **T**[17]). *Suppose that* $x > 0$, *that*

$$u_n = u_n(x) := \alpha e^{-x} \frac{x^{\alpha x+\beta-1}}{\Gamma(\alpha n+\beta)} \quad (n = N, N+1, \ldots),$$

and that δ, γ, ζ, *and* η *satisfy*

$$0 < \delta < \frac{1}{\alpha}, \quad \gamma = \frac{(\alpha\delta)^2}{3}, \quad \frac{1}{2} < \zeta < \frac{2}{3}, \quad 0 < \eta < 2\zeta - 1.$$

Then

(a) $\displaystyle\sum_{n=N}^{\infty} u_n \to 1 \quad as \quad x \to \infty,$

(b) $u_n \leq u_{n+1}$ *if* $n \leq \dfrac{x}{\alpha} - \dfrac{\beta}{\alpha} - 1$, *and*

$u_n \geq u_{n+1}$ *if* $n \geq \dfrac{x}{\alpha} - \dfrac{1-\beta}{\alpha} - 1$,

(c) $\displaystyle\sum_{|n-x/\alpha|>\delta x} u_n = \mathbf{O}\left(e^{-\gamma x}\right),$

(d) $\displaystyle\sum_{|n-x/\alpha|>x^\zeta} u_n = \mathbf{O}\left(e^{-x^\eta}\right),$

(e) $u_n = \dfrac{\alpha}{\sqrt{2\pi x}} e^{-\alpha^2(n-x/\alpha)^2/2x} \left\{1 + \mathbf{O}\left(x^{3\zeta-2}\right)\right\}$ *if* $\left|n - \dfrac{x}{\alpha}\right| \leq x^\zeta$.

Proof

Part (a). This is one of the conditions for the regularity of (B, α, β) and was first given by Borwein [40], page 130.

Part (b). Since

$$\frac{u_{n+1}}{u_n} = \frac{x^\alpha \Gamma(\alpha n + \beta)}{\Gamma(\alpha n + \beta + \alpha)},$$

the results follow from lemma 7.2 with $\nu = \alpha, y = \alpha n + \beta + \alpha$.
Part (c). Let n_1 and n_2 be the integers such that

$$n_1 > \frac{x}{\alpha} + \delta x \geq n_1 + 1 \quad \text{and} \quad n_2 < \frac{x}{\alpha} - \delta x \leq n_2 + 1 .$$

Using Stirling's formula, in the form

$$\Gamma(\alpha n + \beta) = (2\pi)^{1/2} e^{-\alpha n} (\alpha n)^{\alpha n + \beta - 1/2} \left\{ 1 + \mathbf{O}\left(\frac{1}{n}\right) \right\},$$

we get

$$\begin{aligned} u_{n_1} &= \mathbf{O}\left(\frac{e^{-x} x^{\alpha n_1 + \beta - 1}}{e^{-\alpha n_1} (\alpha n_1)^{\alpha n_1 + \beta - 1/2}} \right) \\ &= \mathbf{O}\left(e^{\alpha n_1 - x} x^{-1/2} \left\{ \frac{x}{\alpha n_1} \right\}^{\alpha n_1 + \beta - 1/2} \right) \\ &= \mathbf{O}\left(e^{\alpha \delta x} \left\{ \frac{x}{\alpha n_1} \right\}^{\alpha n_1} \right) = \mathbf{O}\left(e^{\alpha \delta x - \alpha n_1 \log(\alpha n_1 / x)} \right) \\ &= \mathbf{O}\left(e^{\alpha \delta x - (x + \alpha \delta x) \log(1 + \alpha \delta)} \right) = \mathbf{O}\left(e^{-\Delta_1 x} \right), \end{aligned}$$

where

$$\begin{aligned} \Delta_1 &= -\alpha\delta + (1 + \alpha\delta) \log(1 + \alpha\delta) \\ &= \frac{(\alpha\delta)^2}{1 \cdot 2} - \frac{(\alpha\delta)^3}{2 \cdot 3} + \frac{(\alpha\delta)^4}{3 \cdot 4} - \ldots > \frac{1}{3} (\alpha\delta)^2 . \end{aligned}$$

Similarly

$$u_{n_2} = \mathbf{O}\left(e^{-\Delta_2 x} \right),$$

where

$$\begin{aligned} \Delta_2 &= \alpha\delta + (1 - \alpha\delta) \log(1 - \alpha\delta) \\ &= \frac{(\alpha\delta)^2}{1 \cdot 2} + \frac{(\alpha\delta)^3}{2 \cdot 3} + \frac{(\alpha\delta)^4}{3 \cdot 4} + \ldots > \frac{1}{2} (\alpha\delta)^2 . \end{aligned}$$

Next, for $r \geq 0, x \geq 2(1 - \beta)/\alpha\delta$, we have, by lemma 7.2 (ii) with $\nu = \alpha r, y = \alpha n_1 + \beta + \alpha r$,

$$\frac{u_{n_1+r}}{u_{n_1}} = \frac{x^{\alpha r}\Gamma(\alpha n_1 + \beta)}{\Gamma(\alpha n_1 + \beta + \alpha r)} \leq \left(1 + \frac{\alpha\delta}{2}\right)^{-\alpha r},$$

since $0 < x(1 + \alpha\delta/2) \leq \alpha n_1 + \beta - 1$. It follows that

$$\sum_{n - x/\alpha > \delta x} u_n = \sum_{r=0}^{\infty} u_{n_1+r} \leq \sum_{r=0}^{\infty} \left(1 + \frac{\alpha\delta}{2}\right)^{-r}$$
$$= \mathbf{O}\left(e^{-\Delta_1 x}\right) = \mathbf{O}\left(e^{-\gamma x}\right).$$

Finally, by part (b), we have

$$\sum_{n - x/\alpha < -\delta x} u_n = \sum_{n < x/\alpha - \delta x} u_n = \mathbf{O}\left(x e^{-\Delta_2 x}\right) = \mathbf{O}\left(e^{-\gamma x}\right).$$

This completes the proof of part (c).
Part (e). Let $H = n - x/\alpha$, so that $|h| \leq x^{\zeta}$. Using Stirling's formula, we have

$$\log\Gamma(\alpha n + \beta)$$
$$= \frac{1}{2}\log(2\pi) - \alpha n + \left(\alpha n + \beta - \frac{1}{2}\right)\log(\alpha n) + \mathbf{O}\left(\frac{1}{n}\right)$$
$$= \frac{1}{2}\log(2\pi) - x - \alpha h + \left(\alpha h + x + \beta - \frac{1}{2}\right)\log(\alpha h + x) + \mathbf{O}\left(\frac{1}{x}\right)$$
$$= \frac{1}{2}\log(2\pi) - x - \alpha h + \left(\alpha n + \beta - \frac{1}{2}\right)\log(x)$$
$$+ \left(\alpha h + x + \beta - \frac{1}{2}\right)\left\{\frac{\alpha h}{x} - \frac{\alpha^2 h^2}{2x^2} + \mathbf{O}\left(\frac{|h|^3}{x^3}\right)\right\} + \mathbf{O}\left(\frac{1}{x}\right)$$
$$= \frac{1}{2}\log(2\pi) - x - \alpha h + \left(\alpha n + \beta - \frac{1}{2}\right)\log(x)$$
$$+ \alpha h + \frac{a^2 h^2}{2x} + \mathbf{O}\left(\frac{1}{x}\right) + \mathbf{O}\left(\frac{|h|}{x}\right) + \mathbf{O}\left(\frac{|h|^3}{x^2}\right)$$
$$= \frac{1}{2}\log(2\pi) - x - \alpha h + \left(\alpha n + \beta - \frac{1}{2}\right)\log(x)$$
$$+ \frac{\alpha^2 h^2}{2x} + \mathbf{O}\left(x^{3\zeta - 2}\right),$$

since $\frac{1}{2} < \zeta < \frac{2}{3}$ and $|h| \leq x^\zeta$.

Therefore

$$\log(u_n) = \log(\alpha) - x - (\alpha n + \beta - 1)\log(x) - \log(\Gamma(\alpha n + \beta))$$
$$= \frac{1}{2}\log\left(\frac{\alpha^2}{2\pi x}\right) - \frac{\alpha^2 h^2}{2x} + \mathbf{O}\left(x^{3\zeta-2}\right),$$

and so

$$u_n = \frac{\alpha}{\sqrt{2\pi x}} e^{-\alpha^2(n-x/\alpha)^2/2x}\left\{1 + \mathbf{O}\left(x^{3\zeta-2}\right)\right\},$$

as required.

Part (d). Since $e^{-\gamma x} = \mathbf{O}\left(e^{-x^\eta}\right)$, it is sufficient, in view of part (c), to prove that

$$\sum_{\delta x \geq |n-x/a| > x^\zeta} u_n = \mathbf{O}\left(e^{-x^\eta}\right).$$

By parts (b) and (e), the largest term in this sum is $\mathbf{O}\left(e^{-\alpha^2 x^{2\zeta-1/2}}\right)$, and this is sufficient to give the required estimate.

7.1.2 Results on Cesàro sums

First, we recall Hardy's notation for the binomial coefficients as given in chapter 2:

$$E_k^\lambda = \binom{k+\lambda}{k}.$$

Cesàro methods are described briefly in chapter 2. But here we need some properties of Cesàro sums s_n^λ expressed in terms of the series $\sum a_n$. We recall that these are given by the formula

$$s_n^\lambda = \sum_{k=0}^{n} E_k^\lambda\, a_{n-k},$$

so that $s_n^{-1} = a_n$, $s_n^0 = s_n = \sum_{k=0}^{n} a_k$, and

$$s_n^{\lambda+\delta} = \sum_{k=0}^{n} E_k^{\delta-1}\, s_{n-k}^\lambda.$$

Lemma 7.4. (see Borwein [43], lemma 3. Compare Hardy **T**[17], theorem 146). *Suppose that $k > 0$, that*

$$\phi_k(x) := \alpha^k \sum_{n=0}^{\infty} \frac{\Gamma(n+k)}{\Gamma(\alpha n+k)} \frac{x^{\alpha n}}{n!}, \tag{7.1}$$

that $a_{\alpha,\beta}(t)$ is convergent for all $t > 0$, and that $a_n = 0$ for $n < N$. Then, for $x > 0$,

$$\alpha^k \sum_{n=N}^{\infty} s_n^k \frac{x^{\alpha n+\beta+k-1}}{\Gamma(\alpha n+\beta+k)}$$
$$= \frac{1}{\Gamma(k)} \int_0^x (x-t)^{k-1} \phi_k(x-t)\, dt \sum_{n=N}^{\infty} s_n \frac{t^{\alpha n+\beta-1}}{\Gamma(\alpha n+\beta)}. \tag{7.2}$$

Proof. The convergence of $a_{\alpha,\beta}(t)$ for all positive t is equivalent to the convergence of $s_{\alpha,\beta}(t)$ for all positive t. Thus the right hand side of (7.2) is equal to

$$\frac{\alpha^k}{\Gamma(k)} \int_0^x (x-t)^{k-1}\, dt \sum_{n=N}^{\infty} \frac{s_n t^{\alpha n+\beta-1}}{\Gamma(\alpha n+\beta)} \sum_{m=0}^{\infty} \frac{\Gamma(m+k)}{\Gamma(\alpha m+k)} \frac{(x-t)^{\alpha m}}{m!}$$
$$= \frac{\alpha^k}{\Gamma(k)} \sum_{n=N}^{\infty} \frac{s_n}{\Gamma(\alpha n+\beta)} \sum_{m=0}^{\infty} \frac{\Gamma(m+k)}{\Gamma(\alpha m+k)\, m!} \int_0^x t^{\alpha n+b-1} (x-t)^{\alpha m+k-1}\, dt$$
$$= \frac{\alpha^k}{\Gamma(k)} \sum_{n=N}^{\infty} s_n \sum_{m=0}^{\infty} \frac{\Gamma(m+k)\, x^{\alpha n+\alpha m+\beta+k-1}}{m!\, \Gamma(\alpha n+\alpha m+\beta+k)}$$
$$= \alpha^k \sum_{n=N}^{\infty} s_n \sum_{m=N}^{\infty} \binom{m-n+k-1}{m-n} \frac{x^{\alpha m+\beta+k-1}}{\Gamma(\alpha m+\beta+k)}$$
$$= \alpha^k \sum_{n=N}^{\infty} \frac{x^{\alpha m+\beta+k-1}}{\Gamma(\alpha m+\beta+k)} \sum_{m=N}^{\infty} \binom{m-n+k-1}{m-n} s_n$$
$$= \alpha^k \sum_{n=N}^{\infty} \frac{x^{\alpha m+\beta+k-1}}{\Gamma(\alpha m+\beta+k)} s_m^k,$$

as required.

Lemma 7.5. (see Borwein [43], lemma 4). *If* $k \geq 0$ *and if* $\sum_0^\infty a_n = \sigma\,(B, \alpha, \beta)$, *then*

$$\Gamma(k+1)\alpha^{k+1}e^{-x}\sum_{n=N}^{\infty} s_n^k \frac{x^{\alpha n+\beta-1}}{\Gamma(\alpha n+\beta+k)} \to \sigma \tag{7.3}$$

as $x \to \infty$.

Proof. If $k = 0$, there is nothing to prove. Suppose therefore that $k > 0$.

If we knew that $\sum_{n=0}^{\infty} s_n^k$ was summable by the Cesàro method (C, k) to σ, that is, if

$$s_n^k \sim \frac{n^k\,\sigma}{\Gamma(k+1)} \quad \text{as} \quad n \to \infty\,,$$

then we would have

$$\frac{\alpha^k s_n^k\,\Gamma(k+1)}{\Gamma(\alpha n+\beta+k)} \sim \frac{\sigma}{\Gamma(\alpha n+\beta)} \quad \text{as} \quad n \to \infty\,,$$

and (7.3) would follow from the regularity of the (B, α, β) summability method.

Therefore, there is no loss in generality in assuming that

$$a_n = 0 \quad \text{for} \quad n < N\,.$$

Thus, by lemma 7.4, is sufficient to prove that

$$kx^{-k}\int_0^x (x-t)^{k-1}\phi_k(x-t)e^{-(x-t)}s_{\alpha,\beta}(t)\,dt \to \sigma \quad \text{as} \quad x \to \infty\,. \tag{7.4}$$

Since

$$\frac{\alpha^k\,\Gamma(n+k)}{\Gamma(\alpha n+k)\,n!} \sim \frac{\alpha}{\Gamma(an+1)} \quad \text{as} \quad n \to \infty\,,$$

and since the $(B, \alpha, 1)$ summability method is regular, it follows that

$$e^{-x}\phi_k(x) \to 1 \quad \text{as} \quad x \to \infty\,.$$

This means that it is now easy to show that $kx^{-k}t^{k-1}e^{-t}\phi_k(t)$ is the kernel of a regular integral transformation, and since we are given that $s_{\alpha,\beta}(x) \to \sigma$ as $x \to \infty$, the proof of the lemma is complete.

Lemma 7.6. (see Borwein [43], lemma 5). *If* $\sum_{n=0}^{\infty} a_n = 0\,(B, \alpha, \beta)$ *and if*

$$s_n^{k-\mu} = \mathbf{o}\left(n^{\lambda}\right) \qquad (k \le 0,\ 0 < \mu \le 1,\ \lambda > -1,\ \lambda + \mu > 0)\,, \tag{7.5}$$

then

$$s_n^k = \mathbf{o}\left(n^k\right) + \mathbf{o}\left(n^{\lambda+\mu/2}\right).$$

Proof. It follows from (7.5) and a known result proved above (lemma 2.15 in chapter 2), or in Hardy's book (**T**[17], theorem 144), that

$$s_n^k = \mathbf{o}\left(n^{\lambda+\mu}\right), \tag{7.6}$$

and that, if $0 < H < \dfrac{1}{\alpha}$ and $\left|n - \dfrac{x}{\alpha}\right| < Hx$, then

$$s_n^k - s_{[x/\alpha]}^k = \mathbf{o}\left\{\left(|n - x/\alpha|^{\mu} + 1\right) x^{\lambda}\right\} \tag{7.7}$$

uniformly as $x \to \infty$.

Suppose that ζ satisfies $\dfrac{1}{2} < \zeta < \dfrac{2}{3}$, and write

$$\begin{aligned} &\alpha e^{-x} \sum_{n=N}^{\infty} \left(s_n^k - s_{[x/\alpha]}^k\right) \frac{x^{\alpha n+\beta-1}}{\Gamma(\alpha n + \beta + k)} \\ &= \alpha e^{-x} \left(\sum_{N \le n < x/\alpha - x^{\zeta}} + \sum_{x/\alpha - x^{\zeta} \le n \le x/\alpha + x^{\zeta}} + \sum_{n > x/\alpha + x^{\zeta}} \right) \\ &= S_1 + S_2 + S_3\,. \end{aligned}$$

Then, by lemma 7.5

$$\begin{aligned} &S_1 + S_2 + S_3 + \alpha e^{-x} s_{[x/\alpha]}^k \sum_{n=N}^{\infty} \frac{x^{\alpha n+\beta-1}}{\Gamma(\alpha n + \beta + k)} \\ &= \alpha e^{-x} \sum_{n=N}^{\infty} s_n^k \frac{x^{\alpha n+\beta-1}}{\Gamma(\alpha n + \beta + k)} \\ &= \mathbf{o}(1) \quad \text{as} \quad x \to \infty\,, \end{aligned}$$

and hence we have

$$S_1 + S_2 + S_3 + x^{-k} s^k_{[x/\alpha]} \left(1 + \mathbf{o}(1)\right) = \mathbf{o}(1) \quad \text{as} \quad x \to \infty \,. \tag{7.8}$$

Next, by (7.6) and lemma 7.3 (d),

$$\begin{aligned}
S_1 + S_3 &= \mathbf{O}\left(e^{-x} \sum_{N \le n < x/\alpha - x^\zeta} x^{\lambda+\mu} \frac{x^{\alpha n + \beta - 1}}{\Gamma(\alpha n + \beta + k)} \right) \\
&\quad + \mathbf{O}\left(e^{-x} \sum_{n > x/\alpha + x^\zeta} n^{\lambda+\mu} \frac{x^{\alpha n + \beta - 1}}{\Gamma(\alpha n + \beta + k)} \right) \\
&= \mathbf{O}\left(x^{\lambda+\mu-k} e^{-x} \sum_{N \le n < x/\alpha - x^\zeta} \frac{x^{\alpha n + \beta + k - 1}}{\Gamma(\alpha n + \beta + k)} \right) \\
&\quad + \mathbf{O}\left(x^{\lambda+\mu-k} e^{-x} \sum_{n > x/\alpha + x^\zeta} \frac{x^{\alpha n + \beta + k - \lambda - \mu - 1}}{\Gamma(\alpha n + \beta + k - \lambda - \mu)} \right) \\
&= \mathbf{O}\left(x^{\lambda+\mu-k} e^{-x^\eta} \right) \qquad (0 < \eta < 2\zeta - 1) \\
&= \mathbf{o}(1) \quad \text{as} \quad x \to \infty \,.
\end{aligned} \tag{7.9}$$

Further, by (7.7) and lemma 7.3 (d),

$$\begin{aligned}
S_2 &= \mathbf{o}\left(x^{\lambda-k} e^{-x} \sum_{x/\alpha - x^\zeta \le n \le x/\alpha + x^\zeta} \left(|n - x/\alpha|^\mu + 1 \right) \frac{x^{\alpha n + \beta + k - 1}}{\Gamma(\alpha n + \beta + k)} \right) \\
&= \mathbf{o}\left(x^{\lambda-k} \sum_{|h_n| \le x^\zeta} \left(|h_n|^\mu + 1 \right) \frac{\alpha}{\sqrt{2\pi x}} e^{-\alpha^2 {h_n}^2/2x} \right) \qquad (h_n = n - x/\alpha) \\
&= \mathbf{o}\left(x^{\lambda-k-1/2} \int_{-\infty}^{\infty} \left(|t|^\mu + 1 \right) e^{-\alpha^2 t^t/2x} \, dt \right) \\
&= \mathbf{o}\left(x^{\lambda-k+\mu/2} \right) + \mathbf{o}\left(x^{\lambda-k} \right) \\
&= \mathbf{o}\left(x^{\lambda-k+\mu/2} \right) \quad \text{as} \quad x \to \infty \,.
\end{aligned} \tag{7.10}$$

It follows now from (7.8), (7.9), and (7.10) that

$$s^k_{[x/\alpha]}\left(1+\mathbf{o}(1)\right)=\mathbf{o}\left(x^k\right)+\mathbf{o}\left(x^{\lambda+\mu/2}\right)$$

as $x\to\infty$, and this completes the proof.

7.1.3 Proof of the 'o' theorem

Proof of theorem 7.1 Suppose, without loss of generality, that $\sigma=0$. By hypothesis (i), we have that (7.5) holds with $k=0,\mu=1$, and $\lambda=\rho$. Hence by lemma 7.6, we have

$$s_n=s^0_n=\mathbf{o}\left(n^{\rho+1/2}\right), \tag{7.11}$$

since $\rho+\frac{1}{2}\geq 0$.

Suppose that $2\rho+1=m\mu$, where m is an integer and $0<\mu\leq 1$. We shall prove that

$$s^{r\mu}_n=\mathbf{o}\left(n^{\rho+1/2+r\mu/2}\right) \tag{7.12}$$

for $r=0,1,\ldots,m$. From (7.11), we see that (7.12) holds for $r=0$. We proceed by induction: assume that (7.12) holds for some $r<m$, so that (7.5) holds with

$$k=(r+1)\mu,\quad \lambda=\rho+\frac{1}{2}+\frac{r\mu}{2}.$$

Since

$$\frac{1}{2}(r+1)\mu\leq\frac{m\mu}{2}=\rho+\frac{1}{2},$$

it follows from lemma 7.6, that

$$s^{(r+1)\mu}_n=\mathbf{o}\left(n^{(r+1)\mu}\right)+\mathbf{o}\left(n^{\rho+1/2+(r+1)\mu/2}\right)=\mathbf{o}\left(n^{\rho+1/2+(r+1)\mu/2}\right),$$

which is (7.12) with $r+1$ replacing r.

Thus (7.12) holds for $r=0,1,\ldots,m$; and so, in particular, the case $r=m$ gives

$$s^{2\rho+1}_n=\mathbf{o}\left(n^{2\rho+1}\right),$$

so that $\sum_{n=0}^{\infty}a_n$ is summable $(C,2\rho+1)$ to σ. This completes the proof of theorem 7.1.

7.2 The 'O' theorem

In this section, we prove a generalization of theorem 3.20 below. Before proceeding to the actual proof of theorem 7.7, we shall require some preliminary lemmas.

Theorem 7.7. (see Borwein and Robinson [49]). *If*

(i) $a_n = \mathbf{O}(n^{-\frac{1}{2}})$, *and*

(ii) $\sum_{n=0}^{\infty} a_n = \sigma\,(B, \alpha, \beta)$,

then $\sum_{n=0}^{\infty} a_n = \sigma$.

The case $\alpha = \beta = 1$ of this theorem is known and is stated in chapter 3 above (theorem 3.20). Using this, and theorem 5.13 in chapter 5 above, we can immediately obtain a proof of theorem 7.7, valid for $\alpha \geq 1$. However, Borwein and Robinson's proof, given here, makes no appeal to either α or to β, and so is valid for any given α and β.

7.2.1 Preliminary results

Lemma 7.8. (see Borwein and Robinson [49], lemma 1; see also lemma 7.3 above). *Suppose that* $x > 0$, $h = n - \dfrac{x}{\alpha}$, $\dfrac{1}{2} < \zeta < \dfrac{2}{3}$, *and* $0 < \eta < 2\zeta - 1$. *Then, for* $n \geq N$,

$$e^{-x} \sum_{|h| > x^{\zeta}} \frac{x^{\alpha n + \beta - 1}}{\Gamma(\alpha n + \beta)} = \mathbf{O}\left(e^{-x^{\eta}}\right) \tag{7.13}$$

and

$$e^{-x} \frac{x^{\alpha x + \beta - 1}}{\Gamma(\alpha n + \beta)} = \frac{e^{-\alpha^2 h^2 / 2x}}{\sqrt{2\pi x}} \left\{1 + \mathbf{O}\left(\frac{|h| + 1}{x}\right) + \mathbf{O}\left(\frac{|h|^3}{x^2}\right)\right\} \tag{7.14}$$

if $|h| \leq x^{\zeta}$. *Result* (7.14) *also holds with* $h = n - \left[\dfrac{x}{\alpha}\right]$.

Proof. Note that (7.13) is the same as part (d) of lemma 7.3 above. By observing the techniques used in part (e) of lemma 7.3 above, we see that (7.14) would have been obtained if the following two estimates had not been used in the simplification near the end of the proof:

$$\frac{|h|+1}{x} = \mathbf{O}\left(x^{3\zeta-2}\right) \quad \text{and} \quad \frac{|h|^3}{x^2} = \mathbf{O}\left(x^{3\zeta-2}\right).$$

Note that it is necessary to write $\mathbf{O}\left(\frac{|h|+1}{x}\right)$ instead of $\mathbf{O}\left(\frac{|h|}{x}\right)$ in order to deal with the case $h = 0$. Further, $h = n - \left[\frac{x}{\alpha}\right]$ only requires minor modifications of the proof.

Lemma 7.9. (see Borwein and Robinson [49], lemma 2). *For $n = N$, $N+1, \ldots,$*

$$\frac{n^{1/2}}{\Gamma(\alpha n+\beta)} = \mathbf{O}\left(\frac{1}{\Gamma(\alpha n+\beta-1/2)}\right),$$

Proof. This is easily obtained from Stirling's formula:

$$\Gamma(\alpha n+\beta) = \sqrt{2\pi}e^{\alpha n}(\alpha n)^{\alpha n+\beta-1/2}\left\{1+\mathbf{O}\left(\frac{1}{n}\right)\right\}.$$

Lemma 7.10. (see Borwein and Robinson [49], lemma 3). *Suppose that $\frac{1}{2} < \zeta < \frac{2}{3}$. If $a_n = \mathbf{o}(1)$ and $|h| \le n^{\zeta}$, then*

$$s_{n+h} - s_n = \mathbf{o}\left(|h|\right)$$

as $n \to \infty$, uniformly for $|h| \le n^{\zeta}$.

Proof. This is a special case of lemma 2.15.

Lemma 7.11. (see Borwein and Robinson [49], lemma 4). *Suppose that $\frac{1}{2} < \zeta < \frac{2}{3}$. If $s_n = \mathbf{o}\left(n^{1/2}\right)$ and $|h| \le (\alpha n)^{\zeta}$, then*

$$s_{n+h} = \mathbf{o}\left(n^{1/2}\right)$$

as $n \to \infty$, uniformly for $|h| \le n^{\zeta}$.

Proof. Note that

$$\frac{|s_{n+h}|}{n^{1/2}} = \left(1+\frac{h}{n}\right)^{1/2}\frac{|s_{n+h}|}{(n+h)^{1/2}} \le \left(1+\frac{(\alpha n)^{\zeta}}{n}\right)^{1/2}\frac{|s_{n+h}|}{(n+h)^{1/2}}.$$

Since $n+h \ge n-(\alpha n)^{\zeta}$ and $n-(\alpha n)^{\zeta} \to \infty$, as $n \to \infty$, we have

$\dfrac{s_{n+h}}{(n+h)^{1/2}} = \mathbf{o}(1)$ as $n \to \infty$, uniformly for $|h| \le (\alpha n)^\zeta$.

Since $\left(1 + \dfrac{(\alpha n)^\zeta}{n}\right)^{1/2} \to 1$ as $n \to \infty$, the result follows at once.

Lemma 7.12. (see Borwein and Robinson [49], lemma 5). *Suppose that* $\frac{1}{2} < \zeta < \frac{2}{3}$. *If* $s_n = \mathbf{o}\left(n^{1/2}\right)$ *and* $|h| > (\alpha n)^\zeta > 0$, *then*

$$s_{n+h} = \mathbf{o}\left(|h|\right)$$

as $n \to \infty$.

Proof. From $s_n = \mathbf{o}\left(n^{1/2}\right)$, we obtain that $s_{n+h} = \mathbf{o}\left((n+|h|)^{1/2}\right)$. Note that $\frac{3}{2} < \frac{1}{\zeta} < 2$ and $\frac{3}{4} < \frac{1}{2\zeta} < 1$.

Since $|h| > (\alpha n)^\zeta$, we have that $n+|h| = \mathbf{O}\left(|h|^{1/\zeta}\right)+|h| = \mathbf{O}\left(|h|^{1/\zeta}\right)$, so that $(n+|h|)^{1/2} = \mathbf{O}\left(|h|^{1/2\zeta}\right) = \mathbf{O}\left(|h|\right)$, and so the result follows.

Lemma 7.13. (see Borwein and Robinson [49], lemma 6). *Suppose that* n *is a positive integer and that* $a_n = \mathbf{O}\left(n^{-1/2}\right)$, *then, for* $|h| \le n^\zeta$, $s_{n+h} - s_n = \mathbf{O}\left(|h| n^{-1/2}\right)$ *as* $n \to \infty$.

Proof. This is easily verified.

7.2.2 Estimates of some sums as integrals

The next four lemmas give estimates of certain sums in terms of integrals. The proofs are elementary.

In all these lemmas, we assume that n is a positive integer and that c is positive.

Lemma 7.14. (see Borwein and Robinson [49], lemma 7; compare theorem 140 in Hardy **T**[17])

$$\left| \sum_{h=-\infty}^{\infty} e^{-ch^2/n} - 2\int_0^\infty e^{-ct^2/n}\,dt \right| < 1\,.$$

Further,

$$\sqrt{\frac{c}{\pi n}} \sum_{h=-\infty}^{\infty} e^{-ch^2/n} = 1 + \mathbf{O}\left(n^{-1/2}\right), \tag{7.15}$$

uniformly in any finite interval $0 \le c \le k$.

Lemma 7.15. (see Borwein and Robinson [49], lemma 8)

$$\sum_{h=-\infty}^{\infty} |h| e^{-ch^2/n} \geq 2\int_0^{\infty} te^{-ct^2/n}\,dt - \left(\frac{2n}{ec}\right)^{1/2} = \left(\frac{n}{c}\right) - \left(\frac{2n}{ec}\right)^{1/2}.$$

For n sufficiently large,

$$\sum_{h=-\infty}^{\infty} |h| e^{-ch^2/n} \leq 2\int_0^{\infty} te^{-ct^2/n}\,dt + \left(\frac{2n}{ec}\right)^{1/2} = \left(\frac{n}{c}\right) + \left(\frac{2n}{ec}\right)^{1/2}.$$

Further,

$$\sum_{h=-\infty}^{\infty} |h| e^{-ch^2/n} = \mathbf{O}\left(\int_0^{\infty} te^{-ct^2/n}\,dt\right). \tag{7.16}$$

Lemma 7.16. (see Borwein and Robinson [49], lemma 9). *For* $\frac{1}{2} < \zeta < \frac{2}{3}$,

$$\sum_{|h|>(\alpha n)^{\zeta}}^{\infty} |h| e^{-ch^2/n} = \mathbf{O}\left(\int_{(\alpha n)^{\zeta}}^{\infty} te^{-ct^2/n}\,dt\right).$$

Lemma 7.17. (see Borwein and Robinson [49], lemma 10)

$$\begin{aligned}\sum_{h=-\infty}^{\infty} |h|^3 e^{-ch^2/n} &\geq 2\left(\int_0^{\infty} t^3 e^{-ct^2/n}\,dt - \left(\frac{3n}{2ec}\right)^{3/2}\right)\\ &= \frac{n^2}{c^2} - \frac{1}{\sqrt{2}}\left(\frac{3n}{ec}\right)^{3/2}.\end{aligned}$$

For n sufficiently large,

$$\begin{aligned}\sum_{h=-\infty}^{\infty} |h|^3 e^{-ch^2/n} &\leq 2\left(\int_0^{\infty} t^3 e^{-ct^2/n}\,dt + \left(\frac{3n}{2ec}\right)^{3/2}\right)\\ &= \frac{n^2}{c^2} + \frac{1}{\sqrt{2}}\left(\frac{3n}{ec}\right)^{3/2}.\end{aligned}$$

Further,

$$\sum_{h=-\infty}^{\infty} |h|^3 e^{-ch^2/n} = \mathbf{O}\left(\int_0^{\infty} t^3 e^{-ct^2/n}\,dt\right). \tag{7.17}$$

7.2.3 Results on summability (e, c)

In this section, we suppose that $c > 0$.

Definition 7.18. *If*

$$\sqrt{\frac{c}{\pi n}} \sum_{h=-\infty}^{\infty} s_{n+h} e^{-ch^2/n} \to \sigma$$

as $n \to \infty$, then we say that $\sum_{n=0}^{\infty} a_n$ is summable by the method (e, c) to sum σ, and write

$$\sum_{n=0}^{\infty} a_n = \sigma\,(e, c)\,.$$

See Hardy **T**[17], §9.10. Also note that (7.15) implies that the summability method (e, c) is regular.

Lemma 7.19. (see Borwein and Robinson [49], lemma 11). *Suppose that $a_n = \mathbf{o}(1)$, and that*

$$\textit{either} \quad \text{(i)} \sum_{n=0}^{\infty} a_n = \sigma\,(B, \alpha, \beta) \quad \textit{or} \quad \text{(ii)} \sum_{n=0}^{\infty} a_n = \sigma\,(e, c).$$

Then $s_n = \mathbf{o}\left(n^{1/2}\right)$.

Proof. For case (i) see lemma 7.6 with $k = 0, \mu = 1$, and $\lambda = 0$. For case (ii), see theorem 150 in Hardy **T**[17].
Note In lemmas 7.20, 7.21, 7.22, and 7.23 below, we assume all of the following:

$$h = m - n = m - \frac{x}{\alpha}\,;$$

$$\frac{1}{2} < \zeta < \frac{2}{3}\,;$$

$$s_n = \mathbf{o}\left(n^{1/2}\right) \tag{7.18}$$

Lemma 7.20. (see Borwein and Robinson [49], lemma 12)

$$e^{-\alpha n} \sum_{|h| > (\alpha n)^{\zeta}} s_m \frac{(\alpha n)^{\alpha m + \beta - 1}}{\Gamma(\alpha m + \beta)} = \mathbf{o}(1) \quad \textit{as} \quad n \to \infty\,.$$

Proof. Write $I = e^{-\alpha n} \sum_{|h|>(\alpha n)^\zeta} s_m \frac{(\alpha n)^{\alpha m+\beta-1}}{\Gamma(\alpha m+\beta)}$. By (7.18), we have

$$I = \mathbf{O}\left(e^{-\alpha n} \sum_{|h|>(\alpha n)^\zeta} m^{1/2} \frac{(\alpha n)^{\alpha m+\beta-1}}{\Gamma(\alpha m+\beta)}\right).$$

Further, using lemma 7.9, we obtain that

$$I = \mathbf{O}\left((\alpha n)^{1/2} e^{-\alpha n} \sum_{|h|>(\alpha n)^\zeta} \frac{(\alpha n)^{\alpha m+\beta-3/2}}{\Gamma(\alpha m-\beta-1/2)}\right).$$

Thus, where $0 < \eta < 2\zeta - 1$, using (7.13) with $x = \alpha n$,

$$I = \mathbf{O}\left((\alpha n)^{1/2} e^{-(\alpha n)^\eta}\right) = \mathbf{o}(1)$$

as $n \to \infty$, as required.

Lemma 7.21. (see Borwein and Robinson [49], lemma 13)

$$\sum_{|h|\le(\alpha n)^\zeta} s_{n+h} e^{-\alpha h^2/2n} \mathbf{O}\left(\frac{|h|+1}{n}\right) = \mathbf{o}\left(n^{1/2}\right) \quad as \quad n \to \infty.$$

Proof. Write $I = \sum_{|h|\le(\alpha n)^\zeta} s_{n+h} e^{-\alpha h^2/2n} \mathbf{O}\left(\frac{|h|+1}{n}\right)$.

Using lemma 7.11, we obtain that

$$\begin{aligned} I &= \sum_{|h|\le(\alpha n)^\zeta} e^{-\alpha h^2/2n} \mathbf{o}\left(n^{1/2}\right) \mathbf{O}\left(\frac{|h|+1}{n}\right) \\ &= \mathbf{o}\left(n^{-1/2}\right) \sum_{|h|\le(\alpha n)^\zeta} e^{-\alpha h^2/2n} \mathbf{O}\left(|h|+1\right) \\ &= \mathbf{o}\left(n^{-1/2} \sum_{|h|\le(\alpha n)^\zeta} (|h|+1)\right) e^{-\alpha h^2/2n}. \end{aligned}$$

Using lemma 7.14 and (7.16), we obtain that

$$I = \mathbf{o}\left(n^{-1/2} \int_0^\infty (t+1) e^{-\alpha t^2/2n}\, dt\right)$$

$$= \mathbf{o}\left(n^{-1/2}\left\{\frac{n}{\alpha}+\frac{1}{2}\sqrt{\frac{2\pi n}{\alpha}}\right\}\right) = \mathbf{o}\left(n^{1/2}\right)$$

as $n \to \infty$, as required.

Lemma 7.22. (see Borwein and Robinson [49], lemma 14).

$$\sum_{|h|\le(\alpha n)^{\varsigma}} s_{n+h} e^{-\alpha h^2/2n} \mathbf{O}\left(\frac{|h|^3}{n^n}\right) = \mathbf{o}\left(n^{1/2}\right) \quad \textit{as} \quad n \to \infty.$$

Proof. From lemma 7.11 and using (7.17), we obtain

$$\begin{aligned}\sum_{|h|\le(\alpha n)^{\varsigma}} s_{n+h} e^{-\alpha h^2/2n} \mathbf{O}\left(\frac{|h|^3}{n^n}\right) &= \mathbf{o}\left(n^{-3/2}\sum_{|h|\le(\alpha n)^{\varsigma}} |h|^3 e^{-\alpha h^2/2n}\right) \\ &= \mathbf{o}\left(n^{-3/2}\int_0^{\infty} t^3 e^{-\alpha t^2/2n}\,dt\right) \\ &= \mathbf{o}\left(n^{-3/2}\frac{2n^2}{\alpha^2}\right) = \mathbf{o}\left(n^{1/2}\right)\end{aligned}$$

as $n \to \infty$, as required.

Lemma 7.23. (see Borwein and Robinson [49], lemma 15).

$$\sum_{|h|>(\alpha n)^{\varsigma}} s_{n+h} e^{-\alpha h^2/2n} = \mathbf{o}(1) \quad \textit{as} \quad n \to \infty.$$

Proof. Write $I = \sum_{|h|>(\alpha n)^{\varsigma}} s_{n+h} e^{-\alpha h^2/2n}$.

From lemma 7.12, we obtain

$$\begin{aligned} I &= \sum_{|h|>(\alpha n)^{\varsigma}} e^{-\alpha h^2/2n} \mathbf{O}\left(|h|\right) \\ &= \mathbf{O}\left(\sum_{|h|>(\alpha n)^{\varsigma}} |h| e^{-\alpha h^2/2n}\right) \\ &= \mathbf{O}\left(\int_{(\alpha n)^{\varsigma}}^{\infty} t e^{-\alpha t^2/2n}\,dt\right)\end{aligned}$$

$$= \mathbf{O}\left(\frac{n}{\alpha}\, e^{-(\alpha^{2\zeta+1} n^{2\zeta-1})/2}\right)$$

$$= \mathbf{O}\left(n\, e^{-\rho n^{\tau}}\right) = \mathbf{o}(1)$$

as $n \to \infty$, where $\rho = \dfrac{\alpha^{2\zeta+1}}{2}$ and $\tau = 2\zeta - 1 > 0$.

Lemma 7.24. (see Borwein and Robinson [49], lemma 16). *Suppose that k is a positive integer.*

If $\lim_{k\to\infty} s_{\alpha,\beta}(k\alpha) = 0$ *then* $\lim_{x\to\infty} s_{\alpha,\beta}(x) = 0\,.$

Proof. We write $s_{\alpha,\beta}(x) = \alpha e^{-x} \sum_{m=N}^{\infty} s_m \dfrac{x^{\alpha m+\beta-1}}{\Gamma(\alpha m+\beta)}$.

Setting $h = m - n = m - \left[\dfrac{x}{\alpha}\right]$, it follows just as in lemmas 7.20, 7.21, and 7.22, respectively, that

$$e^{-x} \sum_{|h|>x^{\zeta}} s_m \frac{x^{\alpha m+\beta-1}}{\Gamma(\alpha m+\beta)} = \mathbf{o}(1) \quad \text{as} \quad x \to \infty\,,$$

$$\sum_{|h|\le x^{\zeta}} s_{n+h} e^{-\alpha^2 h^2/2x} \mathbf{O}\left(\frac{|h|+1}{x}\right) = \mathbf{o}\left(x^{1/2}\right) \quad \text{as} \quad x \to \infty\,,$$

and

$$\sum_{|h|\le x^{\zeta}} s_{n+h} e^{-\alpha^2 h^2/2x} \mathbf{O}\left(\frac{|h|^3}{x^2}\right) = \mathbf{o}\left(x^{1/2}\right) \quad \text{as} \quad x \to \infty\,.$$

Using these three estimates and (7.14), we obtain that

$$e^{-x} \sum_{m=N}^{\infty} s_m \frac{x^{\alpha m+\beta-1}}{\Gamma(\alpha m+\beta)} = \frac{1}{\sqrt{2\pi x}} e^{-\alpha^2 h^2/2x} s_{n+h} + \mathbf{o}(1) \tag{7.19}$$

as $x \to \infty$.

Setting $x = \alpha n + k$ where $0 \le k < \alpha$, it follows from (7.19) and using lemmas 7.20, 7.21, 7.22, and 7.23, that it is sufficient to prove that

$$\lim_{n\to\infty} (\alpha n)^{-1/2} \sum_{|h|\le x^{\zeta}} s_{n+h} e^{-\alpha h^2/2n} = 0$$

implies that

$$\lim_{x\to\infty} x^{-1/2} \sum_{|h|\le x^\zeta} s_{n+h} e^{-\alpha h^2/2x} = 0\,.$$

Since

$$x^{-1/2} = (\alpha n)^{-1/2} + \mathbf{O}\left(n^{-3/2}\right)$$

and

$$\begin{aligned} e^{-\alpha^2 h^2/a(\alpha n+k)} &= e^{-\alpha^2 h^2/2n}\left\{1+\mathbf{O}\left(\frac{h^2}{n^2}\right)\right\} \\ &= e^{-\alpha^2 h^2/2n}\left\{1+\mathbf{O}\left(\frac{|h|+1}{n}\right)\right\}, \end{aligned}$$

the result follows.

Theorem 7.25. (see Borwein and Robinson [49], theorem 2; compare theorem 151 in Hardy **T**[17]). *Suppose that* $s_n = \mathbf{o}\left(n^{1/2}\right)$. *Then*

$$\sum_{n=0}^{\infty} a_n = \sigma\,(B,\alpha,\beta) \quad \textit{if and only if} \quad \sum_{n=0}^{\infty} a_n = \sigma\left(e, \frac{\alpha}{2}\right).$$

Proof. Since both summability methods are regular, we may assume without loss of generality, that $\sigma = 0$. We may further assume, without loss of generality, that

$$a_0 = a_1 = \ldots = a_{N-1} = 0\,,$$

so that $s_m = 0$ if $m < N$.

Let $x = \alpha n$, $\dfrac{1}{2} < \zeta < \dfrac{2}{3}$, and $h = m - n = m - \dfrac{x}{\alpha}$. Then, using lemma 7.20 and (7.14) with $x = \alpha n$, we obtain

$$\begin{aligned} & e^{-\alpha n} \sum_{m=N}^{\infty} s_m \frac{(\alpha n)^{\alpha m+\beta-1}}{\Gamma(\alpha m+\beta)} \\ &= e^{-\alpha n} \sum_{|h|\le(\alpha n)^\zeta} s_m \frac{(\alpha n)^{\alpha m+\beta-1}}{\Gamma(\alpha m+\beta)} + e^{-\alpha n} \sum_{|h|>(\alpha n)^\zeta} s_m \frac{(\alpha n)^{\alpha m+\beta-1}}{\Gamma(\alpha m+\beta)} \\ &= n^{-1/2}\left\{ \sum_{|h|\le(\alpha n)^\zeta} s_{n+h} e^{-\alpha^2 h^2/2x} \right. \end{aligned}$$

$$+\sum_{|h|\le(\alpha n)^{\varsigma}} s_{n+h}e^{-\alpha^2h^2/2x}\mathbf{O}\left(\frac{|h|+1}{n}\right)$$

$$\left.+\sum_{|h|\le(\alpha n)^{\varsigma}} s_{n+h}e^{-\alpha^2h^2/2x}\mathbf{O}\left(\frac{|h|^3}{n^2}\right)\right\}+\mathbf{o}(1)\,.$$

It now follows, by lemmas 7.21 and 7.22, that

$$e^{-\alpha n}\sum_{m=N}^{\infty} s_m\frac{(\alpha n)^{\alpha m+\beta-1}}{\Gamma(\alpha m+\beta)} = n^{-1/2}\sum_{|h|\le(\alpha n)^{\varsigma}} s_{n+h}e^{-\alpha^2h^2/2x}+\mathbf{o}(1)\,.$$

Since, by lemma 7.23, we have

$$\sum_{|h|>(\alpha n)^{\varsigma}} s_{n+h}e^{-\alpha^2h^2/2x}=\mathbf{o}(1)$$

it follows that

$$e^{-\alpha n}\sum_{m=N}^{\infty} s_m\frac{(\alpha n)^{\alpha m+\beta-1}}{\Gamma(\alpha m+\beta)}=\mathbf{o}(1)\quad\text{as}\quad n\to\infty$$

if and only if

$$n^{-1/2}\sum_{|h|=-\infty}^{\infty} s_{n+h}e^{-\alpha^2h^2/2x}=\mathbf{o}(1)\quad\text{as}\quad n\to\infty\,;$$

that is, if and only if

$$\sum_{n=0}^{\infty}a_n=0\left(e,\frac{\alpha}{2}\right).$$

In view of lemma 7.24, the proof of the theorem is now complete.

Corollary 7.26. *Suppose that* $a_n=\mathbf{o}(1)$. *Then* $\sum_{n=0}^{\infty}a_n=\sigma(B,\alpha,\beta)$ *if and only if* $\sum_{n=0}^{\infty}a_n=\sigma\left(e,\frac{\alpha}{2}\right)$.

This follows easily using lemma 7.19.

Theorem 7.27. (see Borwein and Robinson [49], theorem 3). *Suppose that* $d \in (0, c)$.

If $s_n = \mathbf{o}\left(n^{1/2}\right)$ *and* $\sum_{n=0}^{\infty} a_n = \sigma(e, c)$, *then* $\sum_{n=0}^{\infty} a_n = \sigma(e, d)$.

If $a_n = \mathbf{o}(1)$ *and* $\sum_{n=0}^{\infty} a_n = \sigma(e, c)$, *then* $\sum_{n=0}^{\infty} a_n = \sigma(e, d)$.

Proof. If we had the hypothesis $a_n = \mathbf{O}(1)$, then we would have theorem 155 in Hardy **T**[17]. The modifications required to convert Hardy's proof are minor.

7.2.4 Two preliminary theorems

Theorem 7.28. (see Borwein and Robinson [49], theorem 4; compare theorem 156 in Hardy **T**[17]). *If* $a_n = \mathbf{O}\left(n^{-1/2}\right)$, *and* $\sum_{n=0}^{\infty} a_n = \sigma(B, \alpha, \beta)$, *then* $s_n = \mathbf{O}(1)$.

Proof. Since $a_n = \mathbf{O}\left(n^{-1/2}\right)$, we have that $a_n = \mathbf{o}(1)$, and further, using lemma 7.19, that $s_n = \mathbf{o}\left(n^{1/2}\right)$. Therefore, from theorem 7.25, $\sum_{n=0}^{\infty} a_n = \sigma\left(e, \frac{\alpha}{2}\right)$. Thus

$$\left(\frac{\alpha}{2\pi n}\right)^{1/2} \sum_{h=-\infty}^{\infty} s_{n+h} e^{-\alpha h^2/2n} = \sigma + \mathbf{o}(1)$$

as $n \to \infty$. Since, by (7.15),

$$\left(\frac{\alpha}{2\pi n}\right)^{1/2} \sum_{h=-\infty}^{\infty} e^{-\alpha h^2/2n} = 1 + \mathbf{o}(1)$$

it follows that

$$s_n \{1 + \mathbf{o}(1)\} = \left(\frac{\alpha}{2\pi n}\right)^{1/2} \sum_{h=-\infty}^{\infty} \{s_{n+h} + (s_n - s_{n+h})\} e^{-\alpha h^2/2n}.$$

Thus,

$$s_n \{1 + \mathbf{o}(1)\} = \sigma + \mathbf{o}(1) + \left(\frac{\alpha}{2\pi n}\right)^{1/2} \sum_{|h| > n^{\varsigma}} (s_n - s_{n+h}) e^{-\alpha h^2/2n}$$

$$+\left(\frac{\alpha}{2\pi n}\right)^{1/2} \sum_{|h|\le n^\zeta} (s_n - s_{n+h})\, e^{-\alpha h^2/2n}. \qquad (7.20)$$

Since $a_n = \mathbf{o}(1)$, it follows that $s_n - s_{n+h} = \mathbf{O}(|h|)$. Thus, by lemma 7.16,

$$\begin{aligned}
\left(\frac{\alpha}{2\pi n}\right)^{1/2} \sum_{|h|> n^\zeta} (s_n - s_{n+h})\, e^{-\alpha h^2/2n} &= \mathbf{O}\left(n^{-1/2} \sum_{|h|> n^\zeta} |h| e^{-\alpha h^2/2n}\right) \\
&= \mathbf{O}\left(n^{-1/2} \int_{n^\zeta}^{\infty} t\, e^{-\alpha t^2/2n}\, dt\right) \\
&= \mathbf{O}\left(n^{1/2}\, e^{-\alpha n^{2\zeta-1}/2}\right) \\
&= \mathbf{o}(1) \qquad (7.21)
\end{aligned}$$

as $n \to \infty$, since $\alpha > 0$ and $2\zeta > 1$.

Also, using lemma 7.13 and (7.16), we have

$$\begin{aligned}
\left(\frac{\alpha}{2\pi n}\right)^{1/2} \sum_{|h|\le n^\zeta} (s_n - s_{n+h})\, e^{-\alpha h^2/2n} &= \mathbf{O}\left(\frac{1}{n} \sum_{|h|\le n^\zeta} |h| e^{-\alpha h^2/2n}\right) \\
&= \mathbf{O}\left(\frac{1}{n} \int_{0}^{\infty} t\, e^{-\alpha t^2/2n}\, dt\right) \\
&= \mathbf{O}(1) \qquad (7.22)
\end{aligned}$$

as $n \to \infty$.

It now follows from (7.20), (7.21), and (7.22) that

$$s_n\left(1 + \mathbf{o}(1)\right) = \sigma + \mathbf{o}(1) + \mathbf{o}(1) + \mathbf{O}(1) = \mathbf{O}(1),$$

and so the proof is complete.

In the proof of the next theorem, we need to make use of Vitali's theorem.

Theorem 7.29. Vitali's theorem. (see, for example, **T**[20], p. 117, **T**[28], p. 168, or **T**[13], p. 150). *Let D be a region of the complex plane and suppose that*

(i) *$\{\phi_n\}$ is a sequence of functions analytic and almost uniformly bounded in D;*

(ii) *there exists a sequence* $\{Z_m\}$ *(of distinct* Z*s) in* D *with at least one limit point* $Z_0 \in D$ *such that* $\lim_{n\to\infty} \phi_n(Z_m)$ *exists for* $m = 1, 2, \ldots$.

Then there exists a function $\phi(z)$ *analytic in* D*, such that* $\{\phi_n(z)\}$ *converges almost uniformly to* $\phi(z)$ *in* D.

Note that (i) ensures that $\lim_{n\to\infty} \phi_n(Z_m)$ is necessarily finite.

Theorem 7.30. (see Borwein and Robinson [49], theorem 5). *If* $\sum_{n=0}^{\infty} a_n = \sigma\,(B, \alpha, \beta)$ *and* $s_n = \mathbf{O}(1)$*, then* $\sum_{n=0}^{\infty} a_n = \sigma\,(e, c)$ *for all positive* c.

Proof. Since $s_n = \mathbf{O}(1)$, we have $s_n = \mathbf{o}\left(n^{1/2}\right)$, and thus, by theorem 7.25, that $\sum_{n=0}^{\infty} a_n = \sigma\left(e, \frac{\alpha}{2}\right)$. Therefore, by theorem 7.27,

$$\sum_{n=0}^{\infty} a_n = \sigma\,(e, c) \quad \text{for} \quad 0 < c \le \frac{\alpha}{2}. \tag{7.23}$$

Now let $z = x + iy$ be a complex number. Choose $y_0 > 0$ and $0 < x_0 < \frac{\alpha}{2}$. Define $D = \{z : x > x_0;\ |y| < y_0\}$ and

$$\phi(z) = \left(\frac{z}{\pi n}\right)^{1/2} \sum_{h=-\infty}^{\infty} s_{n+h} e^{-zh^2/n} = \left(\frac{z}{\pi n}\right)^{1/2} \sum_{h=-n}^{\infty} s_{n+h} e^{-zh^2/n}.$$

Since $\left|s_{n+h} e^{-zh^2/n}\right| \le H\left(e^{-x_0/n}\right)^{h^2}$ for some constant H and for all $z \in D$, it follows that $\sum_{h=-n}^{\infty} s_{n+h} e^{-zh^2/n}$ converges uniformly in D, and further that $\phi_n(z)$ is analytic in D $(n = 1, 2, \ldots)$. In D, we have $|z| \le x\left\{1 + \left(\frac{y_0}{x_0}\right)^2\right\}^{1/2}$, and therefore $\left(\frac{|z|}{\pi n}\right)^{1/2} = \mathbf{O}\left(\sqrt{\frac{x}{\pi n}}\right)$ uniformly for all $z \in D$. Thus, by lemma 7.14,

$$|\phi_n(z)| = \mathbf{O}\left(\sqrt{\frac{x}{\pi n}} \sum_{h=-\infty}^{\infty} e^{-xh^2/n}\right)$$

$$= \mathbf{O}\left\{1+\mathbf{O}\left(\sqrt{\frac{x}{n}}\right)\right\} = \mathbf{O}(1),$$

if x is bounded above (say, if $x_0 \le x \le x_1$).

It now follows that $\{\phi_n(z)\}$ is almost uniformly bounded in D, that is, uniformly bounded on compact subsets of D.

Since, by (7.23), $\phi_n(c) \to \sigma$ as $n \to \infty$ for $x_0 \le c \le \frac{\alpha}{2}$, it follows that the sequence $\{\phi_n\}$ satisfies the hypotheses of Vitali's theorem. Thus we obtain that $\phi_n(c) \to \sigma$ as $n \to \infty$ for all $z \in D$. So, in particular, we obtain that $\sum_{n=0}^{\infty} a_n = \sigma\,(e, c)$ for $x_0 \le c < \infty$. Combining this with (7.23) completes the proof of the theorem.

7.2.5 Proof of the 'O' theorem

It follows from theorems 7.28 and 7.30 that

$$\left(\frac{c}{\pi n}\right)^{1/2} \sum_{h=-\infty}^{\infty} s_{n+h} e^{-ch^2/n} = \sigma + \mathbf{o}(1) \tag{7.24}$$

as $n \to \infty$ for all $c > 0$.

Since $a_n = \mathbf{o}(1)$, we have $s_n - s_{n+h} = \mathbf{O}\left(|h|\right)$, and by replacing $\alpha/2$ by c in (7.21), we have

$$\left(\frac{c}{\pi n}\right)^{1/2} \sum_{|h|>n^{\zeta}} \left(s_n - s_{n+h}\right) e^{-ch^2/n} = \mathbf{o}(1) \tag{7.25}$$

as $n \to \infty$.

Using (7.15), we have, for any fixed positive c

$$\left(\frac{c}{\pi n}\right)^{1/2} \sum_{h=-\infty}^{\infty} e^{-ch^2/n} = 1 + \mathbf{o}(1).$$

It follows, from (7.24) and (7.25), that

$$s_n\left\{1+\mathbf{o}(1)\right\} = \left(\frac{c}{\pi n}\right)^{1/2} \sum_{h=-\infty}^{\infty} \left(s_n - s_{n+h}\right) e^{-ch^2/n}$$
$$+ \left(\frac{c}{\pi n}\right)^{1/2} \sum_{h=-\infty}^{\infty} s_{n+h} e^{-ch^2/n}$$

$$= \left(\frac{c}{\pi n}\right)^{1/2} \sum_{|h| \le n^{\varsigma}} (s_n - s_{n+h})\, e^{-ch^2/n} + \mathbf{o}(1) + \sigma + \mathbf{o}(1)\,.$$

Thus, since $s_n = \mathbf{O}(1)$,

$$s_n - \sigma = \left(\frac{c}{\pi n}\right)^{1/2} \sum_{|h| \le n^{\varsigma}} (s_n - s_{n+h})\, e^{-ch^2/n} + \mathbf{o}(1)\,. \tag{7.26}$$

Now, by lemma 7.13, we have $|s_n - s_{n+h}| \le M|h|/n^{-1/2}$ for some constant M and for all h satisfying $|h| \le n^{\varsigma}$ for $n = 1, 2, \ldots$.

Therefore, by lemma 7.15,

$$\begin{aligned} \left| \left(\frac{c}{\pi n}\right)^{1/2} \sum_{|h| \le n^{\varsigma}} (s_n - s_{n+h})\, e^{-ch^2/n} \right| &\le \left(\frac{M}{n}\right)\left(\frac{c}{\pi}\right)^{1/2} \sum_{|h| \le n^{\varsigma}} |h| e^{-ch^2/n} \\ &\le \left(\frac{M}{n}\right)\left(\frac{c}{\pi}\right)^{1/2} \left\{ \frac{n}{c} + \left(\frac{2n}{ec}\right)^{1/2} \right\} \\ &= \frac{M}{\sqrt{\pi c}} + \mathbf{o}(1) \end{aligned}$$

as $n \to \infty$. It now follows from (7.26) that $|s_n - \sigma| \le \dfrac{M}{\sqrt{\pi c}} + \mathbf{o}(1)$. Thus

$$\limsup_{n \to \infty} |s_n - \sigma| \le \limsup_{n \to \infty} \left(\frac{M}{\sqrt{\pi c}} + \mathbf{o}(1) \right) = \frac{M}{\sqrt{\pi c}}\,.$$

Since this holds for all positive c, we have $\lim_{n \to \infty} s_n = \sigma$. This completes the proof of theorem 7.7.

7.3 Kwee's '**O**' theorem

The '**O**' theorems in Hardy **T**[17] and by Borwein and Robinson [49] give conditions that enable one to deduce convergence from Borel summability, whereas the '**o**' theorems in Hardy **T**[17] and by Borwein [43] give conditions that enable one to deduce Cesàro summability from Borel summability.

The next theorem, which is a '**O**' theorem, due to Kwee [161] gives conditions that enable one to deduce Cesàro summability from Borel summability.

Theorem 7.31. (see Kwee [161], theorem 2). *If*

(i) $\rho \geq -\frac{1}{2}$, $a_n = \mathbf{O}(n^\rho)$, *and*

(ii) $\sum_{n=0}^{\infty} a_n = \sigma\,(B,\alpha,\beta)$,

then $\sum_{n=0}^{\infty} a_n = \sigma\,(C, 2\rho+1)$.

When $\rho = -1/2$, this reduces to theorem 7.7 above.

7.3.1 Preliminary results

Lemma 7.32. (see lemma 7.6 above). *If* $\sum_{n=0}^{\infty} a_n$ *is summable* (B,α,β) *and if*

$$s_n^{k-\mu} = \mathbf{O}\left(n^\lambda\right) \qquad (k \leq 0,\ 0 < \mu \leq 1,\ \lambda > -1,\ \lambda + \mu > 0)\,, (7.5)$$

then

$$s_n^k = \mathbf{O}\left(n^k\right) + \mathbf{O}\left(n^{\lambda+\mu/2}\right). \tag{7.27}$$

See the proof of lemma 7.6 above with '**O**' in place of '**o**'.

Lemma 7.33. (see Kwee [161], lemma 2). *If* $\sum_{n=0}^{\infty} a_n = \sigma\,(B,\alpha,\beta)$ *and if* $a_n = \mathbf{O}\,(n^\rho)$ *for* $\rho \geq -1/2$, *then*

$$s_n^p = \mathbf{O}\left(n^{(2\rho+1+p)/2}\right) \tag{7.28}$$

for $0 \leq p \leq 2\rho+1$.

Proof. Let $p = m\mu$, where m is a nonnegative integer and $0 < \mu \leq 1$. We shall prove by induction that

$$s_n^{r/m} = \mathbf{O}\left(n^{(2\rho+1+r\mu)/2}\right) \tag{7.29}$$

for $r = 0, 1, \ldots, m$.

Take $k = 0$, $\mu = 1$, and $l = \rho$ in lemma 7.6. Then (7.27) gives that $s_n = s_n^0 = \mathbf{O}\left(n^{2\rho+1)/2}\right)$, or (7.29) with $r = 0$.

Assume then that (7.29) holds for some $r < m$. Now use (7.27) with $k = (r+1)\mu$ and $\lambda = (2\rho + 1 + r\mu)/2$ to get

$$s_n^{(r+1)\mu} = \mathbf{O}\left(n^{(r+1)\mu}\right) + \mathbf{O}\left(n^{\{2\rho+1+(r+1)\mu\}/2}\right) = \mathbf{O}\left(n^{\{2\rho+1+(r+1)\mu\}/2}\right),$$

since $(r+1)\mu/2 \leq m\mu/2 \leq (2\rho+1)/2$. Thus (7.29) is true for $r = 0, 1, \ldots, m$. In particular, taking $r = m$ in (7.29), we obtain that (7.28) is true for $0 \leq p \leq 2\rho + 1$.

Lemma 7.34. (see Kwee [161], lemma 3). *If* $\sum_{n=0}^{\infty} a_n = \sigma(B, \alpha, \beta)$ *and if* $a_n = \mathbf{O}(n^\rho)$ *for* $\rho \geq -1/2$, *then for* $|h| \leq \delta n, (0 < \delta < 1)$,

(a) $\dfrac{s_n^{2\rho+1}}{E_n^{2\rho+1}} = \mathbf{O}(1)$,

(b) $\dfrac{s_n^{2\rho+1}}{E_n^{2\rho+1}} - \dfrac{s_{n+h}^{2\rho+1}}{E_{n+h}^{2\rho+1}} = \mathbf{O}\left(|h|n^{-1/2}\right)$, *where* $\rho \geq 0$ *or* $\rho = -1/2$,

(c) $\dfrac{s_n^{2\rho+1}}{E_n^{2\rho+1}} - \dfrac{s_{n+h}^{2\rho+1}}{E_{n+h}^{2\rho+1}} = \mathbf{O}\left(|h|n^{-(2\rho+1)/2}\right)$, *where* $0 > \rho > -1/2$.

Proof. Since

$$\frac{s_n^0}{E_n^0} - \frac{s_{n+h}^0}{E_{n+h}^0} = s_n - s_{n+h} = \mathbf{O}\left(|h|n^{-1/2}\right),$$

it follows that (b) holds for $\rho = -1/2$. For $\rho \geq 0$, we use lemma 7.33 to get $s_n^{2\rho+1} = \mathbf{O}\left(n^{2\rho+1}\right)$, which gives (a). Further, we get $s_n^{2\rho} = \mathbf{O}\left(n^{2\rho-1/2}\right)$. Now, we may assume without loss of generality that $h > 0$. Since

$$\begin{aligned}\frac{s_{n+h}^{2\rho+1}}{E_{n+h}^{2\rho+1}} &= \frac{s_0^{2\rho}}{E_0^{2\rho}} + \frac{s_1^{2\rho}}{E_1^{2\rho}} + \ldots + \frac{s_{n+h}^{2\rho}}{E_{n+h}^{2\rho}} \\ &= \frac{s_n^{2\rho+1}}{E_n^{2\rho+1}} + \frac{s_{n+1}^{2\rho}}{E_{n+1}^{2\rho}} + \ldots + \frac{s_{n+h}^{2\rho}}{E_{n+h}^{2\rho}},\end{aligned}$$

we have

$$\frac{s_n^{2\rho+1}}{E_n^{2\rho+1}} - \frac{s_{n+h}^{2\rho+1}}{E_{n+h}^{2\rho+1}} = s_n^{2\rho+1}\left(\frac{1}{E_n^{2\rho+1}} - \frac{1}{E_{n+h}^{2\rho+1}}\right) - \frac{s_{n+1}^{2\rho} + \ldots + s_{n+h}^{2\rho}}{E_{n+h}^{2\rho+1}}$$

$$= \mathbf{O}\left(hn^{-1}\right) + \mathbf{O}\left(hn^{-1/2}\right) = \mathbf{O}\left(hn^{-1/2}\right),$$

completing the proof of part (b).

Note that $0 > \rho > -1/2$ and $s_n^0 = s_n = \mathbf{O}\left(n^{\rho+1/2}\right)$. By lemma 7.33, and by arguing as in the proof of theorem 144 in Hardy **T**[17] with $\alpha = 0, \beta = 2\rho + 1$ and with '**O**' instead of '**o**', we obtain that

$$s_n^{2\rho+1} - s_{n+h}^{2\rho+1} = \mathbf{O}\left\{\left(hn^{1/2}\right)^{2\rho+1}\right\}.$$

Hence

$$\frac{s_n^{2\rho+1}}{E_n^{2\rho+1}} - \frac{s_{n+h}^{2\rho+1}}{E_{n+h}^{2\rho+1}} = \mathbf{O}\left\{\left(hn^{-1/2}\right)^{2\rho+1}\right\}$$

for $h \leq \delta n, 0 < \delta < 1$, completing the proof of (c).

Lemma 7.35. (see Kwee [161], lemma 5). *If $\rho \geq -1/2, a_n = \mathbf{O}(n^\rho)$ and $\sum_0^\infty a_n$ is summable (B, α, β) to σ, then $\dfrac{s_n^{2\rho+1}}{E_n^{2\rho+1}} \to \sigma\left(e, \dfrac{\alpha}{2}\right)$.*

Proof. From lemma 7.5, we have that

$$\sigma_k(x) := \Gamma(k+1)\alpha^{k+1}e^{-x}\sum_{n=N}^{\infty} s_n^k \frac{x^{\alpha n+\beta-1}}{\Gamma(\alpha n+\beta+k)} \to \sigma$$

as $x \to \infty$. Thus we write

$$\sigma_{2\rho+1}(x) = \Gamma(2\rho+2)\alpha^{2\rho+2}e^{-x}\left(\sum_{N\leq n<(x/\alpha-x^\varsigma} + \sum_{|n-x/\alpha|\leq x^\varsigma} + \sum_{n>x/\alpha+x^\varsigma}\right) s_n^{2\rho+1}\frac{x^{\alpha n+\beta-1}}{\Gamma(\alpha n+\beta+2\rho+1)}$$
$$= S_1 + S_2 + S_3 .$$

Since, by lemma 7.34, $s_n^{2\rho+1} = \mathbf{O}\left(E_n^{2\rho+1}\right)$, we obtain from lemma 7.5 (d) that

$$S_1 = \mathbf{O}\left(e^{-x}\sum_{N\leq n<(x/\alpha-x^\varsigma} s_n^{2\rho+1}\frac{x^{\alpha n+\beta-1}}{\Gamma(\alpha n+\beta+2\rho+1)}\right)$$

$$= \mathbf{O}\left(e^{-x} \sum_{N \le n < (x/\alpha - x^{\zeta}} \frac{x^{\alpha n+\beta-1}}{\Gamma(\alpha n+\beta+2\rho+1)}\right)$$

$$= \mathbf{O}\left(e^{-x^{\eta}}\right) = \mathbf{o}(1)\,, \tag{7.30}$$

and

$$S_1 = \mathbf{O}\left(e^{-x} \sum_{n > x/\alpha + x^{\zeta}} s_n^{2\rho+1} \frac{x^{\alpha n+\beta-1}}{\Gamma(\alpha n+\beta+2\rho+1)}\right)$$

$$= \mathbf{O}\left(e^{-x} \sum_{n > x/\alpha + x^{\zeta}} \frac{x^{\alpha n+\beta-1}}{\Gamma(\alpha n+\beta+2\rho+1)}\right)$$

$$= \mathbf{O}\left(e^{-x^{\eta}}\right) = \mathbf{o}(1)\,. \tag{7.31}$$

Now let $x/\alpha = m$ tend to infinity through integer values, and write $n-m = h$. Using Stirling's formula and lemma 7.5 (e), we obtain, for $|h| \le x^{\zeta}$, $\frac{1}{2} < \zeta < \frac{2}{3}$, that

$$\Gamma(2\rho+2)\alpha^{2\rho+2} E_n^{2\rho+1} e^{-x} \frac{x^{\alpha n+\beta-1}}{\Gamma(\alpha n+\beta+2\rho+1)}$$

$$= \sqrt{\frac{\alpha}{2\pi m}} \exp - \frac{\alpha h^2}{2m} \left\{1 + \mathbf{O}\left(m^{3\zeta-2}\right)\right\}\,.$$

Hence

$$S_2 = \sqrt{\frac{\alpha}{2\pi m}} \sum_{|h| \le x^{\zeta}} \frac{s_{m+h}^{2\rho+1}}{E_{m+h}^{2\rho+1}} \exp\left(-\frac{\alpha h^2}{2m}\right)$$

$$+\mathbf{O}\left\{m^{3\zeta-5/2} \sum_{|h| \le x^{\zeta}} \exp\left(-\frac{\alpha h^2}{2m}\right)\right\}$$

$$= \sqrt{\frac{\alpha}{2\pi m}} \sum_{h=-m}^{\infty} \frac{s_{m+h}^{2\rho+1}}{E_{m+h}^{2\rho+1}} \exp\left(-\frac{\alpha h^2}{2m}\right)$$

$$+\mathbf{O}\left\{m^{-1/2} \int_{x^{\zeta}}^{\infty} \exp\left(-\frac{\alpha t^2}{2m}\, dt\right)\right\}$$

$$+\mathbf{O}\left\{m^{3\zeta-5/2}\int_{-\infty}^{\infty}\exp\left(-\frac{\alpha t^2}{2m}\,dt\right)\right\}$$

$$=\sqrt{\frac{\alpha}{2\pi m}}\sum_{h=-m}^{\infty}\frac{s_{m+h}^{2\rho+1}}{E_{m+h}^{2\rho+1}}\exp\left(-\frac{\alpha h^2}{2m}\right)$$

$$+\mathbf{O}\left\{\int_{x^{\zeta}\sqrt{\alpha/2m}}^{\infty}e^{-u^2}\,du\right\}$$

$$+\mathbf{O}\left\{m^{3\zeta-2}\int_{-\infty}^{\infty}e^{-u^2}\,du\right\}$$

$$=\sqrt{\frac{\alpha}{2\pi m}}\sum_{h=-m}^{\infty}\frac{s_{m+h}^{2\rho+1}}{E_{m+h}^{2\rho+1}}\exp\left(-\frac{\alpha h^2}{2m}\right)+\mathbf{o}(1)\,. \tag{7.32}$$

It now follows from (7.30), (7.31), and (7.32) that

$$\sqrt{\frac{\alpha}{2\pi m}}\sum_{h=-m}^{\infty}\frac{s_{m+h}^{2\rho+1}}{E_{m+h}^{2\rho+1}}\exp\left(-\frac{\alpha h^2}{2m}\right)\to\sigma\,,$$

as $m\to\infty$. In other words, that $\dfrac{s_{m+h}^{2\rho+1}}{E_{m+h}^{2\rho+1}}\to\sigma\left(e,\dfrac{\alpha}{2}\right)$.

We conclude this section with a very simple and easily proved result.

Lemma 7.36. *Suppose that $f(x)$ is a nonnegative function which increases for $0\le x\le y$ and decreases for $x\ge y$. Then, for any integer $n>y+1$,*

$$\sum_{k=0}^{n}f(k)\le\int_0^n f(x)\,dx+f(y)\,.$$

7.3.2 Proof of Kwee's 'O' theorem

Let $c=\gamma+i\delta$ be a complex number and let D be a rectangle defined by $\gamma_0<\gamma<\gamma_1, |\delta|<\delta_0$, where $\delta_0>0, 0<\gamma_0<\alpha/2<\gamma_1$. We write

$$\phi_m(c):=\sqrt{\frac{c}{\pi m}}\sum_{h=-m}^{\infty}\frac{s_{m+h}^{2\rho+1}}{E_{m+h}^{2\rho+1}}\exp\left(-\frac{ch^2}{m}\right)\,.$$

From lemma 7.34 (a), we obtain that $s_n^{2\rho+1}=\mathbf{O}\left(E_n^{2\rho+1}\right)$. Thus, for each m, we have that $\phi_m(c)$ is an analytic function of c, regular in D. Further, we have that

$$\phi_m(c) = \mathbf{O}\left\{\sqrt{\frac{|c|}{m}} \sum_{h=-m}^{\infty} \exp\left(-\frac{\gamma h^2}{m}\right)\right\}$$

$$= \mathbf{O}\left\{\sqrt{\frac{\gamma}{m}} \int_{-\infty}^{\infty} e^{-\gamma t^2/m}\, dt\right\} = \mathbf{O}(1)$$

uniformly for $c \in D$ and all m. Since, by lemma 7.34, we have

$$\frac{s_m^{2\rho+1}}{E_m^{2\rho+1}} - \frac{s_{m-1}^{2\rho+1}}{E_{m-1}^{2\rho+1}} = \mathbf{o}(1)\,,$$

it follows from lemma 7.35 and theorem 155 in Hardy's book **T**[17], that $\phi_m(c) \to \sigma$ for each $c \in (\gamma_0, \alpha/2)$ on the real axis. An application of Vitali's theorem now shows that $\phi_m(c) \to \sigma$ for all $c \in D$. Since γ_0 and γ_1 are arbitrary, we obtain that $\phi_m(c) \to \sigma$ for all $c > 0$.

We now have that

$$\frac{s_m^{2\rho+1}}{E_m^{2\rho+1}} - \sigma = \sqrt{\frac{c}{\pi m}} \sum_{h=-m}^{\infty} \left(\frac{s_m^{2\rho+1}}{E_m^{2\rho+1}} - \frac{s_{m+h}^{2\rho+1}}{E_{m+h}^{2\rho+1}}\right) \exp\left(-\frac{ch^2}{m}\right) + \mathbf{o}(1)$$

$$= \sqrt{\frac{c}{\pi m}} \sum_{|h|\le m/z} \left(\frac{s_m^{2\rho+1}}{E_m^{2\rho+1}} - \frac{s_{m+h}^{2\rho+1}}{E_{m+h}^{2\rho+1}}\right) \exp\left(-\frac{ch^2}{m}\right) + \mathbf{o}(1)$$

$$= S_0 + \mathbf{o}(1)\,,$$

say, where, for $\rho \ge 0$ or $\rho = -1/2$, by means of lemma 7.34 (b) and lemma 7.36 with $f(x) = xe^{-cx^2/m}$ and $x_0 = \sqrt{m/2c}$,

$$|S_0| = \mathbf{O}\left\{\frac{\sqrt{c}}{m} \sum_{|h|\le m/z} e^{-\,ch^2/m}\right\}$$

$$= \mathbf{O}\left\{\frac{\sqrt{c}}{m} \sum_{h=-\infty}^{\infty} e^{-\,ch^2/m}\right\}$$

$$= \mathbf{O}\left\{\frac{\sqrt{c}}{m}\left(\frac{m}{c} + \sqrt{\frac{2m}{ce}}\right)\right\}$$

$$= \mathbf{O}\left(\frac{1}{\sqrt{c}}\right) + \mathbf{o}(1)\,. \tag{7.33}$$

We also have, for $0 > \rho > -1/2$ and $\epsilon > 0$,

$$|S_0| = \sqrt{\frac{c}{m}}\left|\left(\sum_{|h|\le c^{-\epsilon-1/2}m^{1/2}} + \sum_{c^{-\epsilon-1/2}m^{1/2}<|h|\le m^{\varsigma}}\right)\right.$$
$$\left.\times\left(\frac{s_m^{2\rho+1}}{E_m^{2\rho+1}} - \frac{s_{m+h}^{2\rho+1}}{E_{m+h}^{2\rho+1}}\right)\exp\left(-\frac{ch^2}{m}\right)\right|\,.$$

Each term in the first sum is $\mathbf{O}(1)$. By applying lemma 7.34 (c) to the second sum, we get

$$|S_0| = \mathbf{O}\left(\frac{1}{c^{\epsilon}}\right) + \mathbf{O}\left(\sqrt{\frac{c}{m}}\, m^{-\rho-1/2}\sum_{c^{-\epsilon-1/2}m^{1/2}<|h|\le m^{\varsigma}} |h|^{2\rho+1}e^{-ch^2/m}\right).$$

For $|h| > c^{-\epsilon-1/2}m^{1/2}$, we have

$$|h|^{2\rho+1} = |h| \times |h|^{2\rho} < |h|\left(c^{-\epsilon-1/2}m^{1/2}\right)^{2\rho} = c^{-(1+2\epsilon)\rho}|h|\,m^{\rho}\,.$$

Hence

$$|S_0| \le \mathbf{O}\left(\frac{1}{c^{\epsilon}}\right) + \mathbf{O}\left(\frac{c^{-(1+2\epsilon)\rho+1/2}}{m}\sum_{c^{-\epsilon-1/2}m^{1/2}<|h|\le m^{\varsigma}} |h|e^{-ch^2/m}\right)$$
$$\le \mathbf{O}\left(\frac{1}{c^{\epsilon}}\right) + \mathbf{O}\left(\frac{1}{c^{(1+2\epsilon)\rho+1/2}}\right) + \mathbf{o}(1)\,. \tag{7.34}$$

Now choose ϵ so that $(1+2\epsilon)\rho + 1/2 > 0$. Since c is arbitrary, it follows from (7.33) and (7.34) that $\limsup_{m\to\infty}\left|\frac{s_m^{2\rho+1}}{E_m^{2\rho+1}} - \sigma\right| = 0$ and the proof is now complete.

7.3.3 Kwee's 'O' theorem is best possible

Kwee [161] has shown that the order of a_n cannot be improved. We give here his example of a divergent series that 'almost' satisfies the conditions of his theorem.

Let $\{p_n\}$ be a sequence of nondecreasing positive numbers such that $p_n \to \infty$. Using this, a divergent sequence $\{s_n\}$ can be constructed with the properties that $s_n = \mathbf{O}(1)$, that $a_n = \mathbf{O}\left(p_n n^{-1/2}\right)$, and that σ_n is summable (B, α, β).

Assume that $\{n_k\}$ and $\{m_k\}$ are sequences of positive integers such that $\{n_k\}$ is strictly increasing, that $m_k \leq n_k$, $2(n_k + 2m_k) < n_{k+1} (k \geq 1)$, $p_{n_k} \geq 2k$, $n_k^{1/2} \geq 4k$, and that m_k is the smallest integer less than or equal to $(2k)^{-1} n_k^{1/2}$ satisfying $m_k \leq k^{-1} n_k^{1/2}$.

Define

$$s_{n_k+m} = \begin{cases} \sin\left(\frac{\pi m}{m_k}\right), & 0 \leq m \leq 2m_k,\ k = 1, 2, \ldots, \\ 0, & \text{otherwise} \end{cases}$$

Clearly we have that $\{s_n\}$ diverges and that $s_n = \mathbf{O}(1)$. Also, $a_n = \mathbf{O}\left(m_k^{-1}\right)$ for $n_k \leq n \leq n_k + 2m_k$, and $a_n = 0$ otherwise. Since $m_k \geq (2k)^{-1} n_k^{1/2}$ and $p_{n_k} \geq 2k$, we obtain that $m_k^{-1} \leq p_{n_k} n_k^{-1/2}$. Further, for $n_k \leq n \leq n_k + 2m_k$, we have

$$p_n n^{-1/2} \geq p_{n_k} \left(n_k + 2m_k\right)^{-1/2} \geq p_{n_k} \left(3n_k\right)^{-1/2},$$

and so, $a_n = \mathbf{O}\left(p_n n^{-1/2}\right)$.

In order to show that $s_n \to 0\,(B, \alpha, \beta)$, it is sufficient to show that, for $\dfrac{1}{2} < \zeta < \dfrac{2}{3}$,

$$\alpha e_{-x} \sum_{|n - x/\alpha| \leq x^\zeta} \frac{s_n x^{\alpha n + \beta - 1}}{\Gamma(\alpha n + \beta)} \to 0 \tag{7.35}$$

as $x \to \infty$. Since $s_n = \mathbf{O}(1)$, it follows that the remainder of the sum defining $s_{\alpha,\beta}(x)$ tends to zero by lemma 7.3 (d).

For x sufficiently large, we have

$$\frac{x}{\alpha} + x^\zeta < 2\left(\frac{x}{\alpha} - x^\zeta\right).$$

Since $2(n_k + 2m_k) < n_{k+1}$, we have that the range of summation in (7.35) cannot overlap more than one of the intervals $[n_k, n_k + 2m_k]$.

If it indeed overlaps none of the intervals, then the sum in (7.35) is zero. On the other hand, if it overlaps one, say $[n_k, n_k + 2m_k]$ for some fixed k, then the number of nonzero terms in the sum does not exceed $2m_k - 1$. For each of these terms, we have $|s_n| \leq 1$, and so

$$\alpha e_{-x} \frac{s_n x^{\alpha n+\beta-1}}{\Gamma(\alpha n+\beta)} \leq M x^{-1/2},$$

where, by lemma 7.3 (b), M is a constant. Thus the sum in (7.35) satisfies

$$\mathbf{O}\left(m_k x^{-1/2}\right) = \mathbf{O}\left(m_k n_k^{-1/2}\right) = \mathbf{O}\left(k^{-1}\right) = \mathbf{o}(1).$$

8

Tauberian theorems – II

This chapter concentrates on the more delicate Tauberian theorems for Borel-type methods.

8.1 The slowly decreasing theorem

For $r \geq 0$ and a real sequence $\{s_n\}$ obtained as the partial sums of the series $\sum_{n=0}^{\infty} a_n$, we consider the following types of Tauberian conditions:

Definition 8.1. Order left
$(L_r) : a_n > -H\,n^{r-1/2}$ *for* $n = 1, 2, \ldots$

Definition 8.2. Slowly decreasing of power order r
$(T_r) : \lim_{\delta \to 0+} \liminf_{n \to \infty} \min_{n \leq m \leq n+\delta\sqrt{n}} \dfrac{s_m - s_n}{n^r} \geq 0.$

Lemma 8.3. (see Borwein and Markovich, [46], page 229). *If* (L_r) *holds then* (T_r) *holds.*

Proof. For $n \leq m \leq n + \delta\sqrt{n}$, we have

$$\frac{s_m - s_n}{n^r} = \frac{1}{n^r} \sum_{j=n+1}^{m} a_j > \frac{-H}{n^r} \sum_{j=n+1}^{m} j^{r-1/2}$$
$$> \frac{-H(m-n)}{\sqrt{n+1}} \left(\frac{m}{n}\right)^r \geq -H\delta \left(1 + \frac{\delta}{\sqrt{n}}\right)^r,$$

so that

$$\lim_{\delta \to 0+} \liminf_{n \to \infty} \min_{n \leq m \leq n+\delta\sqrt{n}} \frac{s_m - s_n}{n^r}$$

$$\geq \lim_{\delta \to 0+} \liminf_{n \to \infty} \left\{ -H\delta \left(1 + \frac{\delta}{\sqrt{n}}\right)^r \right\} = 0 .$$

In this section, we shall have need of the standard Cesàro summability method (C, p), as well as the Valiron method V_α, and the Meyer–König method S_α. For completeness, we reiterate the definitions.

Definition 8.4. *If*

$$C_n^p := A_n^p / E_n^p = \sum_{k=0}^{n} E_{n-k}^{p-1} s_k \Big/ E_n^p \to \sigma$$

as $n \to \infty$, then we say that $s_n \to \sigma\,(C, p)$.

Definition 8.5. *If*

$$\left(\frac{\alpha}{2\pi n}\right)^{1/2} \sum_{k=0}^{\infty} s_k \exp\left\{ -\frac{\alpha(n-k)^2}{2n} \right\} \to \sigma$$

as $n \to \infty$, then we say that $s_n \to \sigma\,(V_\alpha)$.

Definition 8.6. *If $0 < \alpha < 1$ and*

$$(1-\alpha)^{n+1} \sum_{k=0}^{\infty} s_k E_n^k \alpha^k \to \sigma$$

as $n \to \infty$, then we say that $s_n \to \sigma\,(S_\alpha)$.

The two latter methods are needed towards the proof of the main result of this section:

Theorem 8.7. (see Borwein and Markovich [46], theorem 1). *If $s_n \to \sigma\,(B, \alpha, \beta)$ and (T_r) holds, then $s_n \to \sigma\,(C, 2r)$.*

This is the most general tauberian theorem known for Borel-type methods, and includes Kwee's '**O**' theorem [161] Borwein and Robinson's '**O**' theorem [49], and Borwein 's '**o**' theorem [43].

8.1.1 Preliminary results

Lemma 8.8. (see lemma 7.3 above). *Let* $x > 0$,

$$u_n = u_n(x) := \alpha e^{-x} \frac{x^{\alpha x+\beta-1}}{\Gamma(\alpha n+\beta)} \quad (n = N, N+1, \ldots),$$

$h_n = n - x/\alpha$, $\frac{1}{2} < \zeta < \frac{2}{3}$, *and* $0 < \eta < 2\zeta - 1$. *Then*

(i) $\sum_{|h_n|>x^\zeta} u_n(x) = \mathbf{O}\left(e^{-x^\eta}\right)$,

(ii) $u_n(x) = \dfrac{\alpha}{\sqrt{2\pi x}} \exp\left(-\dfrac{\alpha^2 h_n^2}{2x}\right)\left\{1 + \mathbf{O}\left(x^{3\zeta-2}\right)\right\}$

when $|h_n| \leq x^\zeta$.

Lemma 8.9. (compare lemma 7.9 above). *As* $n \to \infty$,

$$\frac{(\alpha n)^r}{\Gamma(\alpha n+\beta)} \sim \frac{1}{\Gamma(\alpha n+\beta-r)}.$$

This follows from Stirling's formula.

Lemma 8.10. (see Borwein and Markovich [46], lemma 4). *Suppose that* $\dfrac{1}{2} < \zeta < \dfrac{2}{3}$, *then as* $x \to \infty$,

(i) $\sum_{N \leq n < x/\alpha - x^\zeta} n^r u_n(x) = \mathbf{o}(1)$,

(ii) $\sum_{n > x/\alpha + x^\zeta} n^r u_n(x) = \mathbf{o}(1)$.

Proof.
Part (i): from lemmas 8.9 and 8.8 (i), we have that, as $x \to \infty$,

$$\begin{aligned}\sum_{N \leq n < x/\alpha - x^\zeta} n^r u_n(x) &= \mathbf{O}\left\{x^r e^{-x} \sum_{N \leq n < x/\alpha - x^\zeta} \frac{x^{\alpha n+\beta-r-1}}{\Gamma(\alpha n+\beta-r)}\right\} \\ &= \mathbf{O}\left(x^r e^{-x^\eta}\right) = \mathbf{o}(1).\end{aligned}$$

Part (ii): the proof is similar.

Lemma 8.11. (see Rajagopal [222], lemma 1). *If $\{s_n\}$ satisfies condition (T_r), then there exist positive constants K, K' such that, for $m \geq n \geq 1$,*

$$s_m - s_n > -K\,m^r\left(m^{1/2} - n^{1/2}\right) - K'\,n^r,$$

$$s_m - s_n \geq -K\left(m^{r+1/2} - n^{r+1/2}\right) - K'\,n^r.$$

This is a simplified form of a more general result stated by Rajagopal in [222]. He does not give a proof, but states that the proof is 'exactly like its case $r = 0$' which is given in Hardy's book **T**[17], theorem 239.

Lemma 8.12. (see Borwein and Markovich [46], lemma 6. Compare Hyslop [125], lemma 1). *Suppose that $h_n = n - x/\alpha, p \geq 0$, and $\frac{1}{2} < \zeta < 1$. Then, as $x \to \infty$,*

$$\text{(i)}\quad \sum_{n > x/\alpha + x^\zeta} n^r h_n^p \exp\left(-\frac{\alpha^2 h_n^2}{2x}\right) = \mathbf{o}(1),$$

$$\text{(ii)}\quad \sum_{0 \leq n < x/\alpha - x^\zeta} n^r |h_n|^p \exp\left(-\frac{\alpha^2 h_n^2}{2x}\right) = \mathbf{o}(1),$$

$$\text{(iii)}\quad \sum_{n=0}^{\infty} n^r |h_n|^p \exp\left(-\frac{\alpha^2 h_n^2}{2x}\right) = \mathbf{O}\left\{x^{r+(p+1)/2}\right\}.$$

Proof. By considering the derivative with respect to n, it is easy to see that

$$(n - x/\alpha)^p \exp\left(-\frac{\alpha^2 (n - x/\alpha)^2}{2x}\right)$$

is increasing or decreasing according as $(x - n/\alpha)$ is greater than or equal to, or less than or equal to, px. Thus

$$\sum_{n > x/\alpha + x^\zeta} n^r (n - x/\alpha)^p \exp\left(-\frac{\alpha^2 (n - x/\alpha)^2}{2x}\right)$$

$$= \mathbf{O}\left\{x^r \sum_{n = x/\alpha + x^\zeta}^{2x/\alpha} (n - x/\alpha)^p \exp\left(-\frac{\alpha^2 (n - x/\alpha)^2}{2x}\right)\right\}$$

$$+\mathbf{O}\left\{\sum_{2x/\alpha}^{\infty}(n-x/\alpha)^{p+r}\exp\left(-\frac{\alpha^2(n-x/\alpha)^2}{2x}\right)\right\}$$

$$=\mathbf{o}(1)+\mathbf{O}\left\{\int_{x/\alpha}^{\infty}t^{p+r}\exp\left(-\frac{\alpha^2(n-x/\alpha)^2}{2x}\right)dt\right\}=\mathbf{o}(1).$$

Part (ii) is similar, and part (iii) follows from considering the finite sum. See lemma 1 in Hyslop [125] for details.

Lemma 8.13. (see Borwein and Markovich [46], lemma 7; compare Sitaraman [257], lemma 5, and Bingham [17], theorem 5). *Let M and L represent any positive integers satisfying*

$$M > x/\alpha + t\sqrt{x/\alpha}, \qquad N < L < x/\alpha - t\sqrt{x/\alpha}.$$

Then, as both t, $x \to \infty$,

(i) $\displaystyle\sum_{n=N}^{L} n^r u_n(x) = \mathbf{o}\left(x^r\right)$,

(ii) $\displaystyle\sum_{n=M}^{\infty} n^r u_n(x) = \mathbf{o}\left(x^r\right)$,

(iii) $\displaystyle\sum_{n=L}^{M} n^r u_n(x) \sim (x/\alpha)^r$,

(iv) $\displaystyle\sum_{n=M}^{\infty}\left(n^{r+1/2}-M^{r+1/2}\right)u_n(x) = \mathbf{o}\left(x^r\right)$.

Note. The precise meaning of part (i), for example, is that given any $\epsilon > 0$, there corresponds an X such that

$$\left|x^{-r}\sum_{n=N}^{L} n^r u_n(x)\right| < \epsilon$$

whenever $x > X, t > X, M > x/\alpha + t\sqrt{x/\alpha}$ and $N < L < x/\alpha - t\sqrt{x/\alpha}$.

This technical lemma is the one that gives the important estimates for the proof of theorem 8.7.

Proof.

Part (i): for $\dfrac{1}{2} < \zeta < \dfrac{2}{3}$, we have

$$0 \le S_1 := \sum_{n=N}^{L} u_n(x) \le \sum_{N \le n \le x/\alpha - t\sqrt{x/\alpha}} u_n(x)$$

$$= \left(\sum_{N \le n \le x/\alpha - x^\zeta} + \sum_{x/\alpha - x^\zeta < n \le x/\alpha - t\sqrt{x/\alpha}} \right) u_n(x)$$

$$:= S_{11} + S_{12}\,, \quad \text{say.}$$

By lemma 8.10 (i), we have that $S_{11} = \mathbf{o}(1)$ as $x \to \infty$. Further, by lemma 8.8 (ii), as t, $x \to \infty$,

$$S_{12} = \mathbf{O}\left\{ x^{-1/2} \sum_{x/\alpha - x^\zeta < x/\alpha - t\sqrt{x/\alpha}} \exp\left(- \frac{\alpha^2 (x/\alpha - n)^2}{2x} \right) \right\}$$

$$= \mathbf{o}(1) + \mathbf{O}\left\{ x^{-1/2} \int_{t\sqrt{x/\alpha}}^{x^\zeta} \exp\left(- \frac{\alpha^2 y^2}{2x} \right) dy \right\}$$

$$= \mathbf{o}(1) + \mathbf{O}\left\{ \int_{t\sqrt{\alpha/2}}^{\infty} \exp\left(- u^2 \right) du \right\} = \mathbf{o}(1)\,.$$

Thus, as t, $x \to \infty$, $S_1 = \mathbf{o}(1)$, and hence

$$0 \le \sum_{n=N}^{L} n^r u_n(x) \le (x/\alpha)^r \sum_{n=N}^{L} u_n(x) = \mathbf{o}\left(x^r\right)\,.$$

Part (ii): for $\frac{1}{2} < \zeta < \frac{2}{3}$, we have

$$S_2 := x^{-r} \sum_{n=M}^{\infty} n^r u_n(x)$$

$$= x^{-r} \left(\sum_{M \le n \le x/\alpha + x^\zeta} + \sum_{n > x/\alpha + t\sqrt{x/\alpha}} \right) n^r u_n(x)$$

$$:= S_{21} + S_{22}, \quad \text{say.}$$

By lemma 8.10 (ii), we have $S_{22} = \mathbf{o}(1)$ as $x \to \infty$. Further, it follows from lemmas 8.9 and 8.8 (ii), that

$$S_{21} = \mathbf{O}\left\{ x^{-r} \sum_{x/\alpha + t\sqrt{x/\alpha} < n \le x/\alpha + x^{\varsigma}} n^r u_n(x) \right\}$$

$$= \mathbf{O}\left\{ e^{-x} \sum_{x/\alpha + t\sqrt{x/\alpha} < n \le x/\alpha + x^{\varsigma}} \frac{x^{\alpha n + \beta - r - 1}}{\Gamma(\alpha n + \beta - r} \right\}$$

$$= \mathbf{O}\left\{ x^{-1/2} \sum_{x/\alpha + t\sqrt{x/\alpha} < n \le x/\alpha + x^{\varsigma}} \exp\left(- \frac{\alpha^2 (n - x/\alpha)^2}{2x} \right) \right\}.$$

By following the method of part (ii), we can deduce that as $x,\, t \to \infty, S_{21} = \mathbf{o}(1)$ and so we can complete the proof of part (ii).

Part (iii): the case $r = 0$ follows from parts (i) and (ii) with $r = 0$ and the known result (part of the regularity of the (B, α, β) method), that $\sum_{n=N}^{\infty} u_n(x) \to 1$ as $x \to \infty$.

In order to prove the result for $r > 0$, we observe that this is equivalent to proving the following:

$$\sum_{n=L_i}^{m_i} n^r\, u_n(x) \sim (x_i/\alpha)^r$$

as $i \to \infty$, *whenever* $\{M_i\}, \{L_i\}, \{t_i\}, \{x_i\}$ *are sequences such that* $t_i \to \infty, x_i \to \infty$, *and*

$$M_i > x_i/\alpha + t_i\sqrt{x_i/\alpha} \quad \textit{and} \quad q < L_i < x_i/\alpha - t_i\sqrt{x_i/\alpha}\,.$$

Suppose therefore that $\{M_i\}$, $\{L_i\}$, $\{t_i\}$, $\{x_i\}$ are sequences satisfying the above conditions, and let

$$w_i = \min\{(x_i)^{1/4}, t_i\}\,,$$

so that

$$0 \le w_i \le t_i, \quad w_i \to \infty, \quad \text{and} \quad w_i/\sqrt{x_i} \to 0.$$

Now choose sequences of positive integers $\{m_{i'}\}$, $\{L_{i'}\}$ such that

$$M_{i'} - 1 \le x_i/\alpha + w_i\sqrt{x_i/\alpha} < M_{i'} \le M_i,$$

$$L_{i'} \le L_{i'} < x_i/\alpha - w_i\sqrt{x_i/\alpha} < L_{i'} + 1.$$

Then

$$\sum_{n=L_i}^{M_i} n^r u_n(x_i) = \left(\sum_{n=L_i}^{L_{i'}-1} + \sum_{n=L_{i'}}^{M_{i'}} + \sum_{n=M_{i'}+1}^{M_i} \right) n^r u_n(x_i). \tag{8.1}$$

(The first series on the right hand side of (8.1) is defined to be zero if $L_{i'} = L_i$, as is the last series if $M_{i'} = M_i$.)
Since

$$(L_{i'})^r \sum_{n=L_{i'}}^{M_{i'}} u_n(x_i) \le \sum_{n=L_{i'}}^{M_{i'}} n^r u_n(x_i) \le (M_{i'})^r \sum_{n=L_{i'}}^{M_{i'}} u_n(x_i)$$

and

$$(L_{i'})^r \sim (x_i/\alpha)^r, \quad (M_{i'})^r \sim (x_i/\alpha)^r \quad \text{as} \quad i \to \infty,$$

it follows that

$$\sum_{n=L_{i'}}^{M_{i'}} n^r u_n(x_i) \sim (x_i/\alpha)^r \sum_{n=L_{i'}}^{M_{i'}} u_n(x_i)$$

$$= (x_i/\alpha)^r \left(\sum_{n=N}^{\infty} - \sum_{n=N}^{L_{i'}-1} - \sum_{n=M_{i'}+1}^{\infty} \right) u_n(x_i)$$

as $i \to \infty$.
Since

$$\sum_{n=N}^{\infty} u_n(x_i) \to 1 \quad \text{as} \quad i \to \infty$$

we have, by parts (i) and (ii) with $r = 0$, that

$$\sum_{n=L_{i'}}^{M_{i'}} n^r u_n(x_i) \sim (x_i/\alpha)^r \quad \text{as} \quad i \to \infty. \tag{8.2}$$

Further, from (8.1) and (8.2), and parts (i) and (ii), we obtain

$$\sum_{n=L_i}^{M_i} n^r u_n(x_i) \sim (x_i/\alpha)^r \quad \text{as} \quad i \to \infty,$$

as required.

Part (iv): an application of the mean value theorem shows that in order to prove the desired result it suffices to show that

$$S_4 := x^{-r} \sum_{n=M}^{\infty} \left(\sqrt{n} - \sqrt{M}\right) n^r u_n(x) = \mathbf{o}(1) \quad \text{as} \quad t, x \to \infty.$$

To prove this observe that since $M > x/\alpha$ we have

$$\frac{(n-M)\sqrt{\alpha/x}}{2} \geq \sqrt{n} - \sqrt{M}$$

and hence

$$\begin{aligned} 0 \leq S_4 &\leq \sqrt{\alpha/2}\, x^{-r-1/2} \sum_{n=M}^{\infty} (n-M) n^r u_n(x) \\ &= \sqrt{\alpha/2}\, x^{-r-1/2} \left\{ \sum_{M \leq n \leq x/\alpha + x^{\zeta}} \right. \\ &\qquad\qquad \left. + \sum_{n > x/\alpha + x^{\zeta}} \right\} (n-M) n^r u_n(x) \\ &=: S_{41} + S_{42} \end{aligned}$$

where $1/2 < \zeta < 2/3$.

Since $M > x/\alpha + t\sqrt{x/\alpha}$ and $n - M < n - x/\alpha$, it follows from lemmas 8.9 and 8.8 (ii) that

$$S_{41} = \mathbf{O}\left\{ x^{-1/2} e^{-x} \sum_{x/\alpha + t\sqrt{x/\alpha} \leq n \leq x/\alpha + x^{\zeta}} (n - x/\alpha) n^r \frac{x^{\alpha n + \beta - r - 1}}{\Gamma(\alpha n + \beta)} \right\}$$

$$
\begin{aligned}
&= \mathbf{O}\left\{ x^{-1/2} e^{-x} \sum_{x/\alpha + t\sqrt{x/\alpha} \le n \le x/\alpha + x^{\varsigma}} (n - x/\alpha) \frac{x^{\alpha n + \beta - r - 1}}{\Gamma(\alpha n + \beta - r} \right\} \\
&= \mathbf{O}\left\{ x^{-1} \sum_{x/\alpha + t\sqrt{x/\alpha} \le n \le x/\alpha + x^{\varsigma}} (n - x/\alpha) \exp\left(- \frac{\alpha^2 (n - x/\alpha)^2}{2x} \right) \right\} \\
&= \mathbf{o}(1) + \mathbf{O}\left\{ x^{-1} \int_{t\sqrt{x/\alpha}}^{\infty} y \exp\left(- \frac{\alpha^2 y^2}{2x} \right) dy \right\} \\
&= \mathbf{o}(1) + \mathbf{O}\left\{ \int_{t\sqrt{\alpha/2}}^{\infty} u \exp(-u^2)\, du \right\} = \mathbf{o}(1) \quad \text{as} \quad t, x \to \infty.
\end{aligned}
$$

Next, by lemmas 8.9 and 8.8 (i), we have that, as $x \to \infty$,

$$
\begin{aligned}
S_{42} &= \mathbf{O}\left\{ x^{1/2} e^{-x} \sum_{n > x/\alpha + x^{\varsigma}} n^{r+1} \frac{x^{\alpha n + \beta - r - 2}}{\Gamma(\alpha n + \beta)} \right\} \\
&= \mathbf{O}\left\{ x^{1/2} e^{-x} \sum_{n > x/\alpha + x^{\varsigma}} n^{r+1} \frac{x^{\alpha n + \beta - r - 2}}{\Gamma(\alpha n + \beta - r - 1)} \right\} \\
&= \mathbf{O}\{ x^{1/2} e^{-x^{\eta}} \} = \mathbf{o}(1).
\end{aligned}
$$

It follows that $S_4 = \mathbf{o}(1)$ as $t, x \to \infty$.

Theorem 8.14. (see Borwein and Markovich [46], theorem 2). *Suppose that $\{s_n\}$ is a sequence such that (T_r) holds and that $s_{\alpha,\beta}(x) = \mathbf{O}(x^r)$ as $x \to \infty$. Then $s_n = \mathbf{O}(n^r)$.*

Proof. We follow the method used by Sitaraman in [257], theorem 1. Define

$$
\begin{aligned}
\sigma_n &:= n^{-r} s_n\,, \\
\sigma_1(n) &:= \max_{\nu \le n} \sigma_\nu\,, \\
\sigma_2(n) &:= \max_{\nu \le n} (-\sigma_\nu)\,.
\end{aligned}
$$

We shall assume that $\{\sigma_n\}$ is unbounded and show that this leads to a contradiction.

There are then two possibilities:

Case (A): $\sigma_1(n) \geq \sigma_2(n)$ for infinitely many values of n.

Case (B): $\sigma_1(n) < \sigma_2(n)$ for all n sufficiently large.

First, suppose that case (A) holds. Then, from the assumption, we have that $\sigma_1(n) \to \infty$. We write

$$\begin{aligned} s_{\alpha,\beta}(x) &= \left(\sum_{n=N}^{L-1} + \sum_{n=L}^{M-1} + \sum_{n=M}^{\infty} \right) u_n(x)\, s_n \\ &:= T_1(x) + T_2(x) + T_3(x), \quad \text{say,} \end{aligned} \tag{8.3}$$

where first L and then M are chosen as follows: corresponding to any positive $H > \sigma_1(N)$, there exist integers $L = L(H) > 0$ such that

$$\sigma_L = \sigma_1(L) > 2H, \quad \sigma_1(L) \geq \sigma_2(L). \tag{8.4}$$

Take the least value of L and then the least $M = M(H) > L$ such that

$$\sigma_M \leq \frac{1}{2}\sigma_L. \tag{8.5}$$

There are such Ms when H is large, for otherwise $\sigma_n \to \infty$, and then lemma 8.9 and the total regularity of the $(B, \alpha, \beta - r)$ method (Hardy **T**[17], theorem 9) would imply that

$$x^{-r} s_{\alpha,\beta}(x) \to \infty \quad \text{as} \quad x \to \infty,$$

contradicting the hypothesis $s_{\alpha,\beta}(x) = \mathbf{O}(x^r)$. In view of lemma 8.11, and the choice of M and L in (8.4) and (8.5), we have that

$$K(M^{1/2} - L^{1/2}) > \sigma_1(N) \left\{ \left(\frac{L}{M} \right)^r - \frac{1}{2} \right\} - K',$$

where K and K' are positive constants (compare Sitaraman [257], proof of theorem 1). Now we have either

$$\left(\frac{L}{M} \right)^r > \frac{3}{4} \quad \text{or} \quad \left(\frac{M}{L} \right)^r \geq \frac{4}{3}.$$

In the first case,

$$K(M^{1/2} - L^{1/2}) > \frac{1}{4}\sigma_1(L) - K',$$

while in the second case

$$M^{1/2} - L^{1/2} \geq L^{1/2}\left\{\left(\frac{4}{3}\right)^{1/(2r)} - 1\right\}.$$

Hence

$$t := t(H) = \frac{1}{2}(M^{1/2} - L^{1/2}) \to \infty \tag{8.6}$$

as $L \to \infty$ (or $H \to \infty$). Next, let

$$x := x(H) = \frac{\alpha}{4}(M^{1/2} + L^{1/2})^2 \tag{8.7}$$

so that $x \to \infty$ as $H \to \infty$, since $M > L \to \infty$. It follows from (8.6) and (8.7) that

$$M > x/\alpha + t\sqrt{x/\alpha}, \quad q < L < x/\alpha - t\sqrt{x/\alpha}, \tag{8.8}$$

where $t, x \to \infty$ as $H \to \infty$.

In the analysis which follows, suppose that L, M, and x are chosen as in (8.4), (8.5), and (8.7) and consequently satisfy (8.9). Therefore $t, x \to \infty$ as $H \to \infty$ and the properties (i), (ii), (iii), and (iv) of lemma 8.13 hold. With reference to (8.3), we see that as $H \to \infty$,

$$T_1(x) \geq -\sigma_2(L)\sum_{n=N}^{L-1} n^r u_n(x) \tag{8.9}$$

$$\geq -\sigma_1(L)\sum_{n=N}^{L} n^r u_n(x) = -\sigma_1(L)\mathbf{o}(1), \tag{8.10}$$

by lemma 8.13 (i). Further, since M is the least integer greater than L which satisfies (8.5), we have

$$\sigma_n > \frac{1}{2}\sigma_L = \frac{1}{2}\sigma_1(L) \quad \text{for} \quad L \leq n \leq M-1. \tag{8.11}$$

Thus, as $H \to \infty$,

$$T_2(x) > \frac{1}{2}\sigma_1(L)\sum_{n=N}^{M-1} n^r u_n(x) \sim \frac{1}{2}\sigma_1(L)(x/\alpha)^r \tag{8.12}$$

by lemma 8.13 (iii).

Next, by lemma 8.11, there are positive constants K and K' such that

$$s_n - s_{M-1} \geq -K(n^{r+1/2} - M - 1)^{r+1/2}) - K'(M-1)^r$$

for $n \geq M$. Thus

$$\begin{aligned} s_n &> s_{M-1} - K(n^{r+1/2} - (M-1)^{r+1/2}) - \mathbf{O}(M^{r-1/2}) - K'M^r \\ &> -K(n^{r+1/2} - M^{r+1/2}) - \mathbf{O}(M^r) \end{aligned} \tag{8.13}$$

for $n \geq M$, since, by (8.11) and (8.4),

$$s_{M-1} = \sigma_{M-1}(M-1)^r > \frac{1}{2}\sigma_L(M-1)^r > H(M-1)^r > 0.$$

By (8.13) and lemma 8.13 (ii) and (iv), we have

$$\begin{aligned} T_3(x) &\geq -K\sum_{n=M}^{\infty}(n^{r+1/2} - M^{r+1/2})u_n(x) - \mathbf{O}(1)\sum_{n=M}^{\infty} n^r u_n(x) \\ &\geq -\mathbf{o}(x^r) \quad \text{as} \quad H \to \infty. \end{aligned} \tag{8.14}$$

Substituting (8.9), (8.12) and (8.14), in (8.3), we get

$$x^{-r}s_{\alpha,\beta}(x) \geq \sigma_1(L)\left(\frac{1}{2}\alpha^{-r} - \mathbf{o}(1)\right) - \mathbf{o}(1) \to \infty \text{ as } H \to \infty,$$

since $\sigma_1(L) \to \infty$ as $L \to \infty$ (or $H \to \infty$). This implies that $x^{-r}s_{\alpha,\beta}(x)$ is unbounded above, contradicting the hypothesis $s_{\alpha,\beta}(x) = \mathbf{O}(x^r)$.

Next, suppose that case (B) holds (that is, there exists an M_0 such that $\sigma_2(n) > \sigma_1(n)$ for $n \geq M_0$). Then, in view of our underlying assumption, we have $\sigma_2(n) \to \infty$. Now write

$$s_{\alpha,\beta}(x) = \left(\sum_{n=N}^{L} + \sum_{n=N+1}^{M} + \sum_{n=M+1}^{\infty}\right) u_n(x)s_n$$

$$:= T_1(x) + T_2(x) + T_3(x)\,, \tag{8.15}$$

where first M and then L are chosen as follows. Corresponding to any positive $H > \sigma_2(M_0)$ choose the least $M = M(H)$ such that

$$\sigma_2(n) > \sigma_1(n) \quad \text{for} \quad n \geq M, \quad \sigma_M = -\sigma_2(M) < -2H. \tag{8.16}$$

Then choose the largest $L = L(H) \in (N, M)$ for which

$$\sigma_L \geq \frac{1}{2}\sigma_M = -\frac{1}{2}\sigma_2(M). \tag{8.17}$$

There are such L's when H is large, for otherwise $\sigma_n \to -\infty$ and then lemma 8.9 and the total regularity of the $(B, \alpha, \beta - r)$ method would imply that

$$x^{-r} s_{\alpha,\beta}(x) \to -\infty \quad \text{as} \quad x \to \infty,$$

contradicting the hypothesis $s_{\alpha,\beta}(x) = \mathbf{O}(x^r)$.

The choices of M and L in (8.15) and (8.16) and lemma 8.11 imply that there are positive constants K, K' for which

$$\begin{aligned} K(M^{1/2} - L^{1/2}) &\geq \sigma_2(M)\left\{1 - \frac{1}{2}\left(\frac{N}{M}\right)^r\right\} - K'\left(\frac{N}{M}\right)^r \\ &\geq \frac{1}{2}\sigma_2(M) - K' \to \infty \end{aligned}$$

as $H \to \infty$ (compare Sitaraman [257], proof of Theorem 1). Hence defining $t = t(H)$ and $x = x(H)$ as in (8.6) and (8.7) we see that $t, x \to \infty$ as $H \to \infty$, and that (8.8) holds. Consequently, as $H \to \infty$, the properties (i), (ii), (iii), and (iv) of lemma 8.13 hold. The rest of the proof of case (B) is exactly as given in Sitaraman ([257], case (ii) of theorem 1) with the roles of L and M interchanged. This rules out the possibility of case (B) holding.

Lemma 8.15. (compare Faulhaber [81], lemma 5). *Suppose $h_n = n - x/\alpha$, $0 < H < 1$, $(1 - H)x/\alpha \leq n \leq (1 + H)x/\alpha$, and k is any integer ≥ 2. Then, as $x \to \infty$*

$$u_n(x) = \frac{\alpha}{\sqrt{2\pi x}} \exp\left(-\frac{\alpha^2 h_n^2}{2x} + g_k + R_k\right),$$

where

$$R_k = \mathbf{O}\left\{\frac{|h_n|^{k+1}+1}{x^k}\right\}, \quad g_k = \sum_{i=1}^{k}\sum_{j=0}^{i+1} b_{i,j}\frac{h_n^j}{x^i},$$

and the b_i, js are constants with $b_{1,2} = b_{k,k+1} = 0$.

Note: In particular, the result is true for all n such that $|h_n| \le x^\zeta$, $1/2 < \zeta < 2/3$.
Proof. Since

$$\alpha n = \alpha h_n + x \quad \text{and} \quad 0 < 1 - H \le \frac{\alpha h_n}{x} + 1 \le 1 + H$$

it follows from a form of Stirling's formula that, as $x \to \infty$,

$$\begin{aligned} &\log\Gamma(\alpha n + \beta) \\ &= (\alpha h_n + x + \beta - 1/2)\log x - \alpha h_n - x + (1/2)\log 2\pi \\ &\quad + (\alpha h_n + x + \beta - 1/2)\log\left(\frac{\alpha h_n}{x} + 1\right) \\ &\quad + \sum_{r=1}^{k}\frac{(-1)^{r+1}B_{r+1}(\beta)}{r(r+1)x^r}\left(\frac{\alpha h_n}{x}+1\right)^{-r} + \mathbf{O}\left(\frac{1}{x^{k+1}}\right), \end{aligned} \tag{8.18}$$

where $k \ge 1$ and each $B_{r+1}(\beta)$ is a Bernoulli polynomial. Since

$$\left|\frac{\alpha h_n}{x}\right| \le H < 1$$

we have

$$\left(\frac{\alpha h_n}{x}+1\right)^{-r} = \sum_{j=0}^{k-r}\binom{j}{-r}\left(\frac{\alpha h_n}{x}\right)^j + \mathbf{O}\left\{\left(\frac{|h_n|}{x}\right)^{k-r+1}\right\}, \tag{8.19}$$

and

$$\log\left(\frac{\alpha h_n}{x}+1\right) = \sum_{j=1}^{k}\frac{(-1)^{j-1}}{j}\left(\frac{\alpha h_n}{x}\right)^j + \mathbf{O}\left\{\left(\frac{|h_n|}{x}\right)^{k+1}\right\}. \tag{8.20}$$

It follows from (8.19) that

$$
\begin{aligned}
&\sum_{r=1}^{k} \frac{(-1)^{r+1} B_{r+1}(\beta)}{r(r+1)x^r} \left(\frac{\alpha h_n}{x}+1\right)^{-r} \\
&= \sum_{r=1}^{k}\sum_{j=0}^{k-r} d_{r,j} \frac{h_n^j}{x^{r+j}} + \sum_{r=1}^{k} \frac{1}{x^r} \mathbf{O}\left\{\left(\frac{|h_n|}{x}\right)^{k-r+1}\right\} \\
&= \sum_{i=1}^{k}\sum_{j=0}^{i-1} d_{i-j,j} \frac{h_n^j}{x^i} + \mathbf{O}\left\{\frac{|h_n|^{k+1}+1}{x^{k+1}}\right\}, \qquad (8.21)
\end{aligned}
$$

where the $d_{r,j}$s are constants.

If we denote the double sum on the right hand side of (8.21) by t_k and then substitute (8.20) and (8.21) in (8.17) we obtain, after some simplification,

$$
\begin{aligned}
&\log u_n(x) \\
&= \log\alpha - x + (\alpha h_n + x + \beta - 1)\log x - \log\Gamma(\alpha n+\beta) \\
&= \log\frac{\alpha}{\sqrt{2\pi x}} + \alpha h_n \\
&\quad + (\alpha h_n + x + \beta - 1/2)\sum_{j=1}^{k} \frac{(-1)^j}{j}\left(\frac{\alpha h_n}{x}\right)^j - t_k \\
&\quad + \mathbf{O}\left\{\frac{|h_n|^{k+1}+1}{x^k}\right\} \quad \text{as} \quad x\to\infty. \qquad (8.22)
\end{aligned}
$$

We now combine the **O**-term with the term

$$\frac{(-1)^k(\alpha h_n)^{k+1}}{kx^k}$$

on the right hand side of (8.21) into R_k to get, after a further simplification,

$$\log u_n(x) = \log\frac{\alpha}{\sqrt{2\pi x}} - \frac{\alpha^2 h_n^2}{2x} + g_k + R_k,$$

where

$$R_k = \mathbf{O}\left\{\frac{|h_n|^{k+1}+1}{x^k}\right\} \text{ and}$$

$$g_k = \sum_{i=1}^{k}\sum_{j=0}^{i+1} b_{i,j}\frac{h_n^j}{x^i},$$

with $b_{1,2} = b_{k,k+1} = 0$.

Lemma 8.16. (see Hardy and Littlewood [112]). *Suppose that ϕ and ψ are increasing functions satisfying the relation $\phi = \mathbf{O}(\psi)$, that $f^{(r)}(x)$ is continuous, and that*

$$f = \mathbf{o}(\phi), \qquad x^r f^{(r)} = \mathbf{O}(\psi).$$

Then, for $0 < r < s$,

$$x^s f^{(s)}(x) = \mathbf{o}\left(\phi^{(r-s)/r}\psi^{s/r}\right).$$

The proof of this result is quite long. Since the result is not part of the development of Borel's methods, it is not given here.

8.2 An equivalence theorem

Lemma 8.17. (see Hyslop [125], lemma 3 or Faulhaber [81], lemma 3). *Suppose that $s_n = \mathbf{O}(n^r)$, and that*

$$\sum_{n=0}^{\infty} s_n \exp\left(-\frac{\alpha(n-x)^2}{2x}\right) = \mathbf{o}\left(x^{b+1/2}\right)$$

as $x \to \infty$, where $b \geq 0$. Then, for each integer $j \geq 0$ and for each $\epsilon > 0$,

$$\sum_{n=0}^{\infty} s_n (n-x)^j \exp\left(-\frac{\alpha(n-x)^2}{2x}\right) = \mathbf{o}\left(x^{b+(j+1)/2+\epsilon}\right)$$

as $x \to \infty$.

Proof. Choose $\epsilon > 0$ and let

$$p = [2r+j], \quad r = \max\left(p, \left[\frac{rj}{\epsilon}\right]+1\right).$$

If $f(x)$ denotes the function

$$\sum_{n=0}^{\infty} s_n \exp\left(-\frac{\alpha(n-x)^2}{2x}\right),$$

then it is easy to verify that $f^{(k)}(x)$ is of the form

$$\sum_{n=0}^{\infty} \exp\left(-\frac{\alpha(n-x)^2}{2x}\right)$$
$$\times\left\{\sum_{\nu=0}^{k} a_{\nu,k}(x-n)^{\nu}x^{-k+[(k-\nu)/2]} + \sum_{\nu=k+1}^{2k} a_{\nu,k}(x-n)^{\nu}x^{-\nu}\right\},$$

where $a_{\nu,k}$ is a polynomial in x^{-1} and $a_{k,k}$ has a nonzero constant term. Using the relation $s_n = \mathbf{O}(n^r)$ and applying lemma 8.12 to each term in this sum, we obtain

$$f^{(k)}(x) = \mathbf{O}\left(x^{r-(k-1)/2}\right).$$

By hypothesis, we have that $f(x) = \mathbf{o}\left(x^{b+1/2}\right)$. Thus, by lemma 8.16, we obtain that

$$f^{(s)}(x) = \mathbf{O}\left(x^{b-(s-1)/2+(rs)/k}\right) \quad (0 < r < s).$$

If we define

$$\phi_s(x) := \sum_{n=0}^{\infty} s_n\,(x-n)^s \exp\left(-\frac{\alpha(n-x)^2}{2x}\right),$$

then we have the following relations:

$$\sum_{\nu=1}^{s} a_{\nu,s}x^{-s+[(s-\nu)/2]}\phi_{\nu} + \sum_{\nu=s+1}^{2s} a_{\nu,s}x^{-\nu} = \mathbf{o}\left(x^{b-(s-1)/2+(rs/k)}\right),$$

where $s = 1, 2, \ldots, p$. Because of the choice of p and k, certain terms may be discarded from these equations. Thus we can rewrite them as

$$\sum_{\nu=1}^{s} a_{\nu,s}x^{-s+[(s-\nu)/2]}\phi_{\nu} + \sum_{\nu=s+1}^{t} a_{\nu,s}x^{-\nu} = \mathbf{o}\left(x^{-(\gamma-1)/2+\epsilon}\right),$$

where $t = \max(p, 2s)$ and

$$\gamma = \begin{cases} s & s \leq j\,, \\ j & s > j\,. \end{cases}$$

We now have p equations in the p unknowns $\phi_1, \phi_2, \ldots, \phi_p$. By considering the relevant determinant, we obtain the result.

Lemma 8.18. (see Hyslop [125], theorem 2 or Faulhaber [81], lemma 4 with $q = 0$). *Suppose that $s_n = \mathbf{O}\left(n^r\right), h_n = n - x/\alpha$, and that*

$$\sum_{k=0}^{\infty} s_k \exp\left(-\frac{\alpha(n-k)^2}{2n}\right) = \mathbf{o}\left(n^{1/2}\right)$$

as $n \to \infty$. Then

$$\sum n = 0^{\infty} s_n \exp\left(-\frac{\alpha^2 h_n^2}{2x}\right) = \mathbf{o}\left(x^{-1/2}\right)$$

as $x \to \infty$.

The proof, as given in Hyslop [125], is similar in construction to the proof of lemma 8.17 given above, and is omitted.

Theorem 8.19. (see Borwein and Markovich [46], theorem 3; compare Hyslop [125], theorems 3 and 6). *Suppose that $s_n = \mathbf{O}(n^r)$. Then $s_n \to \sigma(B, \alpha, \beta)$ if and only if $s_n \to \sigma(V_\alpha)$.*

Proof. Let

$$1/2 < \zeta < 2/3, \quad h_n = n - x/\alpha,$$

$$\overline{s_{\alpha,\beta}}(x) := \sum_{|h_n| \leq x^{\zeta}} u_n(x) s_n \quad \text{and}$$

$$\overline{t}(x) := \frac{\alpha}{\sqrt{2\pi x}} \sum_{|h_n| \leq x^{\zeta}} \exp\left(-\frac{\alpha^2 h_n^2}{2x}\right) s_n.$$

We first prove that $s_n \to \sigma(V_\alpha)$ implies $s_n \to \sigma(B, \alpha, \beta)$. Because of the regularity of both methods it suffices to prove this result for $\sigma = 0$. Suppose therefore that $s_n \to 0(V_\alpha)$. In order to show that $s_n \to 0(B, \alpha, \beta)$ it is enough, by lemma 8.10, to prove that $\overline{s_{\alpha,\beta}}(x) = \mathbf{o}(1)$ as $x \to \infty$. By lemma 8.15, for x sufficiently large and an integer $k > 2r + 1$, we have

$$\begin{aligned}
&\overline{s_{\alpha,\beta}}(x) - \overline{t}(x) \\
&= \sum_{|h_n|\le x^{\zeta}} s_n \left\{ u_n(x) - \exp\left(-\frac{\alpha^2 h_n^2}{2x}\right)\right\} \\
&= \frac{\alpha}{\sqrt{2\pi x}} \sum_{|h_n|\le x^{\zeta}} s_n \exp\left(-\frac{\alpha^2 h_n^2}{2x}\right) \sum_{\mu=1}^{\infty} \frac{(g_k+R_k)^{\mu}}{\mu!} \\
&= \frac{\alpha}{\sqrt{2\pi x}} (A_1(x) + A_2(x) + A_3(x)), \qquad (8.23)
\end{aligned}$$

where

$$A_1(x) := \sum_{|h_n|\le x^{\zeta}} s_n \exp\left(-\frac{\alpha^2 h_n^2}{2x}\right) \sum_{\mu=1}^{2s} \frac{g_k^{\mu}}{\mu!}, \tag{8.24}$$

$$A_2(x) := \sum_{|h_n|\le x^{\zeta}} s_n \exp\left(-\frac{\alpha^2 h_n^2}{2x}\right) \sum_{\mu=1}^{2s} \frac{(g_k+R_k)^{\mu} - g_k^{\mu}}{\mu!}, \tag{8.25}$$

$$A_3(x) := \sum_{|h_n|\le x^{\zeta}} s_n \exp\left(-\frac{\alpha^2 h_n^2}{2x}\right) \sum_{\mu=2s+1}^{2s} \frac{(g_k+R_k)^{\mu}}{\mu!}, \tag{8.26}$$

and the integer $s > r - 1/2$.

We proceed to show that each of the above is $\mathbf{o}(x^{1/2})$ as $x \to \infty$.

To see that $A_1(x) = \mathbf{o}(x^{1/2})$ as $x \to \infty$, consider, for $1 \le \mu \le 2s$,

$$v(x) := \sum_{|h_n|\le x^{\zeta}} s_n \exp\left(-\frac{\alpha^2 h_n^2}{2x}\right) \frac{g_k^{\mu}}{\mu!}.$$

The expansion of g_k given in lemma 8.15 shows that g_k^{μ} is a finite combination of terms of the form $x^{-i} h_n^j$, where

(i) $0 \le j \le \mu$ for $i = \mu$, and

(ii) $0 \le j \le i + \mu$ for $i \ge \mu + 1$.

Hence, if we can show that

$$v_{i,j} := \sum_{|h_n|\le x^{\zeta}} s_n \exp\left(-\frac{\alpha^2 h_n^2}{2x}\right) \frac{h_n^j}{x^i} = \mathbf{o}(x^{1/2})$$

as $x \to \infty$ for the is and js in (i) and (ii), it will follow that

$$v(x) = \mathbf{o}(x^{1/2}),$$

and hence that

$$A_1(x) = \mathbf{o}(x^{1/2})$$

as $x \to \infty$. Now our hypotheses, together with lemma 8.18, and lemma 8.17 with $b = 0$, $\epsilon = 1/4$, imply that, for each integer $j \geq 0$,

$$\sum_{n=0}^{\infty} s_n h_n^j \exp\left(-\frac{\alpha^2 h_n^2}{2x}\right) = \mathbf{o}(x^{(j+1)/2+1/4})$$

as $x \to \infty$. An application of lemma 8.12 shows that, for each integer $j \geq 0$,

$$v_{i,j} = \mathbf{o}(x^{-i+j/2+3/4})$$

as $x \to \infty$. From this it is clear that, in both cases (i) and (ii), $v_{i,j} = \mathbf{o}(x^{1/2})$ and hence that

$$A_1(x) = \mathbf{o}(x^{1/2})$$

as $x \to \infty$.

To prove that $A_2(x) = \mathbf{o}(x^{1/2})$ as $x \to \infty$, it suffices to show that, for $1 \leq \mu \leq 2s$,

$$u(x) := \sum_{|h_n| \leq x^\zeta} n^r \exp\left(-\frac{\alpha^2 h_n^2}{2x}\right) |(g_k + R_k)^\mu - g_k^\mu| = \mathbf{o}(x^{1/2})$$

as $x \to \infty$. Since $k \geq 2$, $1/2 < \zeta < 2/3$, and $|h_n| \leq x^\zeta$ we have, by lemma 8.15, that, as $x \to \infty$,

$$R_k = \mathbf{O}\left\{\frac{|h_n|^{k+1} + 1}{x^k}\right\} = \mathbf{O}(1) \quad \text{and} \quad g_k = \mathbf{O}(1).$$

Hence,

$$\begin{aligned} |(g_k + R_k)^\mu - g_k^\mu| &\leq \sum_{j=1}^{\mu} \binom{\mu}{j} |R_k|^j |g_k|^{\mu-j} \\ &= \mathbf{O}(|R_k|) = \mathbf{O}\left\{\frac{|h_n|^{k+1} + 1}{x^k}\right\}, \end{aligned}$$

and so,

$$u(x) = \mathbf{O}\left\{x^{-k} \sum_{|h_n| \le x^\zeta} n^r(1 + |h_n|^{k+1}) \exp\left(-\frac{\alpha^2 h_n^2}{2x}\right)\right\}$$

as $x \to \infty$. By lemma 8.12, since $k > 2r + 1$,

$$u(x) = \mathbf{O}(x^{r-k+1/2}) + \mathbf{O}(x^{r-k/2+1}) = \mathbf{o}(x^{1/2})$$

as $x \to \infty$.

Finally, to show that $A_3(x) = \mathbf{o}(x^{1/2})$ as $x \to \infty$, we observe that, since $1/2 < \zeta < 2/3$, and $|h_n| \le x^\zeta$, we have, by lemma 8.15,

$$g_k + R_k = g_2 + R_2 = \mathbf{O}\left\{\frac{|h_n| + 1}{x} + \frac{|h_n|^3}{x^2}\right\}. \tag{8.27}$$

In particular, $g_k + R_k = \mathbf{o}(1)$ as $x \to \infty$ and hence

$$\sum_{\mu=2s+1}^{\infty} \frac{|g_k + R_k|^\mu}{\mu!} = \mathbf{O}\left(|g_k + R_k|^{2s+1}\right)$$

as $x \to \infty$. Thus, from this and (8.27), we obtain

$$\begin{aligned}
A_3(x) &= \mathbf{O}\left\{\sum_{|h_n| \le x^\zeta} n^r \exp\left(-\frac{\alpha^2 h_n^2}{2x}\right) \sum_{\mu=2s+1}^{\infty} \frac{|g_k + R_k|^\mu}{\mu!}\right\} \\
&= \mathbf{O}\left\{\sum_{|h_n| \le x^\zeta} n^r \exp\left(-\frac{\alpha^2 h_n^2}{2x}\right) |g_k + R_k|^{2s+1}\right\} \\
&= \mathbf{O}\left\{\sum_{|h_n| \le x^\zeta} n^r \exp\left(-\frac{\alpha^2 h_n^2}{2x}\right) \left(\frac{1 + |h_n|^{2s+1}}{x^{2s+1}} + \frac{|h_n|^{6s+3}}{x^{4s+2}}\right)\right\}.
\end{aligned}$$

Hence, by lemma 8.12, since $x > r - 1/2$,

$$A_3(x) = \mathbf{O}(x^{-2s+r-1/2}) + \mathbf{O}(x^{-s+r}) + \mathbf{O}(x^{-s+r}) = \mathbf{o}(x^{1/2})$$

as $x \to \infty$.

Consequently, it follows from (8.23) that

$$\overline{s_{\alpha,\beta)}}(x) - \bar{t}(x) = \mathbf{o}(1)$$

as $x \to \infty$. Next, by our hypotheses, lemma 8.18, and lemma 8.12 with $p = 0$, we have that $\overline{t}(x) = \mathbf{o}(1)$ as $x \to \infty$. Therefore $\overline{s_{\alpha,\beta}}(x) = \mathbf{o}(1)$ as $x \to \infty$. This completes the proof of the first part of the theorem.

We now prove that $s_n \to \sigma(B,\alpha,\beta)$ implies $s_n \to \sigma(V_\alpha)$. Again it is enough to prove the result for $\sigma = 0$ and we do this by following (Faulhaber [81], theorem II). Suppose that $s_n \to 0(B,\alpha,\beta)$. Then by lemmas 8.10 and 8.15 we have

$$\frac{\alpha}{\sqrt{2\pi x}} \sum_{|h_n| \le x^\varsigma} s_n \exp\left(-\frac{\alpha^2 h_n^2}{2x} + g_k + R_k\right) = \mathbf{o}(1)$$

as $x \to \infty$, that is,

$$\overline{t}(x) + \sqrt{\alpha}\sqrt{2\pi x}(A_1(x) + A_2(x) + A_3(x)) = \mathbf{o}(1)$$

as $x \to \infty$, where A_1, A_2, A_3 are defined by (8.24), (8.25), (8.26), respectively, with $k > 2r + 1$ and the integer $s > r - 1/2$.

Observe that in the proof of the first part of the theorem we only required the hypothesis $s_n = \mathbf{O}(n^r)$ to establish that A_2 and A_3 were $\mathbf{o}(\sqrt{x})$. Since the hypothesis is still operative we now have

$$\overline{t} + \frac{\alpha}{\sqrt{2\pi x}} A_1(x) = \mathbf{o}(1) \tag{8.28}$$

as $x \to \infty$. Further, by lemma 8.12 (iii) with $p = 0$, we have $\overline{t}(x) = \mathbf{O}(x^r)$. Let

$$\gamma := \inf\{\delta : \overline{t}(x) = \mathbf{O}(x^\delta)\}.$$

Then either $\gamma < 0$ or $0 \le \gamma \le r$. We wish to show that $\overline{t}(x) = \mathbf{o}(1)$ as $x \to \infty$ in either case. This is evidently so when $\gamma < 0$. Suppose therefore that $0 \le \gamma \le r$. Consider $A_1(x)$, and for $1 \le \mu \le 2s$ let

$$p(x) := \sum_{|h_n| \le x^\varsigma} s_n \exp\left(-\frac{\alpha^2 h_n^2}{2x}\right) \frac{g_k^\mu}{\mu!},$$

where g_k^μ is a finite combination of terms of the form $x^{-i} h_n^j$ with

(i) $0 \le j \le \mu$ for $i = \mu$, and

(ii) $0 \le j \le i + \mu$ for $i \ge \mu + 1$.

For the is and js in (i) and (ii) let

$$p_{i,j} := \sum_{|h_n| \le x^\varsigma} s_n \exp\left(-\frac{\alpha^2 h_n^2}{2x}\right) \frac{h_n^j}{x^i}.$$

Since $\bar{t}(x) = \mathbf{o}\left(x^{\gamma+1/8}\right)$ as $x \to \infty$ it follows, by lemma 8.12 (i) and (ii) with $p = 0$, that

$$\sum_{n=0}^{\infty} s_n \exp\left(-\frac{\alpha^2 h_n^2}{2x}\right) = \mathbf{o}\left(x^{1/2+\gamma+1/8}\right) \quad \text{as} \quad x \to \infty.$$

Next, it follows from lemma 8.17 with $b = \gamma + 1/8$ and $\epsilon = 1/8$ that, for each integer $j \geq 0$,

$$\sum_{n=0}^{\infty} s_n h_n^j \exp\left(-\frac{\alpha^2 h_n^2}{2x}\right) = \mathbf{o}\left(x^{j/2+\gamma+3/4}\right) \quad \text{as} \quad x \to \infty.$$

Lemma 8.12 implies that, for each integer $j \geq 0$,

$$\sum_{|h_n| \leq x^{\varsigma}} s_n h_n^j \exp\left(-\frac{\alpha^2 h_n^2}{2x}\right) = \mathbf{o}\left(x^{j/2+\gamma+3/4}\right) \quad \text{as} \quad x \to \infty.$$

Thus,

$$p_{i,j} = \mathbf{o}\left(x^{-i+j/2+\gamma+3/4}\right) = \mathbf{o}\left(x^{\gamma+1/4}\right) \quad \text{as} \quad x \to \infty,$$

in both cases (i) and (ii). It follows that

$$A_1(x) = \mathbf{o}\left(x^{\gamma+1/4}\right) \quad \text{as} \quad x \to \infty,$$

and hence, by (8.28), that

$$\bar{t}(x) = \mathbf{o}\left(x^{\gamma-1/4}\right) + \mathbf{o}(1) \quad \text{as} \quad x \to \infty.$$

Now if $\gamma > 1/4$, then

$$\bar{t}(x) = \mathbf{o}\left(x^{\gamma-1/4}\right),$$

and this contradicts the definition of γ. Hence $\gamma \leq 1/4$ and so

$$\bar{t}(x) = \mathbf{o}(1)$$

as $x \to \infty$. It follows, by lemma 8.12 (i) and (ii) with $p = 0$, that

$$\frac{\alpha}{\sqrt{2\pi x}} \sum_{n=0}^{\infty} s_n \exp\left(-\frac{\alpha^2 h_n^2}{2x}\right) = \mathbf{o}(1)$$

as $x \to \infty$, so that $s_n \to \sigma(V_\alpha)$.

8.3 Proof of the slowly decreasing theorem

The hypothesis $s_n \to \sigma(B, \alpha, \beta)$ implies that $s_{\alpha,\beta}(x) = \mathbf{O}(x^r)$ as $x \to \infty$ and hence, by theorem 8.14, that $s_n = \mathbf{O}(n^r)$. Theorem 8.19 now shows that $s_n \to \sigma(V_\alpha)$ while theorem 5.13 shows that there is no loss in generality in making the restriction $0 < \alpha < 1$. It follows by a result due to Faulhaber [81] or Bingham [15] that $s_n \to \sigma(S_{1-\alpha})$ and hence, by a result due to Sitaraman ([257], theorem 2), that $s_n \to \sigma(C_{2r})$.

8.4 Gap theorems

Another kind of Tauberian condition involves series with 'gaps'. These are series in which $\sum_{n=0}^{\infty} a_n$ has sets of consecutive zeros. When such sets of consecutive zeros have certain length properties, it is sometimes possible to deduce the convergence of the given series.

We shall say that a series $\sum_{n=0}^{\infty} a_n$ has a *gap* $(n_k, n_{k'})$, if $a_n = 0$ for $n_n < n < n_{k'}$. We shall also use *gap conditions*, in which it is always assumed that there are infinitely many gaps with $n_{k'} < n_{k+1} \quad k = 0, 1, \ldots$.

Three examples of gap theorems for Borel's method (B) are:

Theorem 8.20. (see, for example, Hardy **T**[17], theorem 141, or Zygmund [319]). *Suppose that $\lambda > 0$. Then there is a number $\delta = \delta(\lambda) > 0$ such that, if*

(i) $n_{k'}/n_k \geq 1 + \lambda > 0$,

(ii) $s_n = \mathbf{O}\left((1+\delta)^n\right)$,

(iii) $s_n \to \sigma(B)$,

then $s_n \to \sigma$;

Theorem 8.21. (see Gaier [87]). *Suppose that the series $\sum_{n=0}^{\infty} a_n$ satisfies*

$$a_n = 0 \quad \textit{for} \quad n \neq n_k, \quad \textit{where} \quad n_{k+1} - n_k \geq \lambda\sqrt{n_k} \quad (\lambda > 0, \ \textit{constant}).$$

If the series is summable (B) or (B'), then the series is convergent;

and finally,

Theorem 8.22. (see, for example, Ingham [128]). *There is a number h, independent of k, such that, if*

(i) $n_{k+1} = n'_k + 1$,

(ii) $\sqrt{n_{k+1}} - \sqrt{n_k} > h > 0$,

(iii) $s_n \to \sigma\,(B)$,

then $s_n \to \sigma$.

The first such gap theorem for the Borel method was given by Pitt [218]; he had the condition $a_n = \mathbf{O}((1+\epsilon)^n)$ (for each $\epsilon > 0$) on the terms of the series. Further to Pitt's theorem, and with the help of Fabry's gap theorem, Meyer–König [186] weakened the condition to $a_n = \mathbf{O}(K^n) K < \infty$. Later, Pitt remarked in his book **T**[22] that his original proof had errors, resulting in Meyer–König and Zeller [187] producing a new proof, this time using functional analysis. In 1956, it was Erdös [78] who considered the rate of growth of a_n, and, amongst other things, used the gap condition $\sum (n_{k+1} - n_k)^{-1} < \infty$. Gaier's theorem imposes no such condition on the a_n, or on the lengths of the gaps.

On the sharpness of Gaier's gap condition, Lorentz [169] has shown that $\sqrt{n_k}$ cannot be reduced to $n^{\frac{1}{2}-\epsilon}$, and Erdös ([77], p. 51) remarked that the condition $n_{k+1} - n_k \geq \epsilon_k \sqrt{n_k}$ (with $\epsilon_k \to 0$) cannot be a gap condition for the Euler method, and so (since $E \to B$), neither for the Borel method.

Krishnan has used Ingham's 'peak-function' method to extend this theorem to Borel-type summability (B, α, β). The reader may find the details in his paper [159].

9

Relationships with other methods

9.1 Product methods with the Cesàro method

We recall that

$$A_n^k = \sum_{r=0}^{n} E_r^k a_{n-r}$$

and that if $t_n^k = \Gamma(k+1)n^{-k}s_n^k \to \sigma$ as $n \to \infty$, then we say that $s_n \to \sigma\,(C,k)$.

Suppose that $f(x)$ is continuous for $x \geq 0$, and that $k > 0$. We recall the Cesàro integral of f:

$$f_k(x) = \frac{1}{\Gamma(k)} \int_0^x (x-t)^{k-1} f(t)\,dt\,.$$

If

$$\Gamma(k+1)x^{-k} f_k(x) \to \sigma$$

as $x \to \infty$, then we say that $f(x) \to \sigma\,(C,k)$.

9.1.1 Product methods

Definition 9.1.
If $t_n^k \to \sigma\,(B,\alpha,\beta)$, then we say that $s_n \to \sigma\,(B,\alpha,\beta)(C,k)$ where $k > -1$. If $S(x) \to \sigma\,(C,k)$, then we say that $s_n \to \sigma\,(C,k)(B,\alpha,\beta)$ where $k > 0$.

We shall prove the following result:

Theorem 9.2. (see Shawyer [250]). *If $k > 0$ and $s_n \to \sigma\,(C,k)(B,\alpha,\beta)$, then $s_n \to \sigma\,(B,\alpha,\beta)(C,k)$.*

With the exception of row-finite methods which commute, such as Hausdorff methods, there are not very many known examples where $P[Q] \subseteq$

$Q[P]$. We shall write $P[Q]$ rather than PQ to indicate the iterated transformation rather than the transformation by the matrix product. Here P is the Cesàro method $C = (C, k)$ and Q is the Borel-type method $B = (B, \alpha, \beta)$.

Since we trivially have $B \subseteq C[B]$, C being regular, and since Borwein [43] has shown that $B \subseteq B[C]$, an example typical of a class of problems considered *inter alia* by Szasz [269], [270] and Jakimovski [137], the above theorem shows that we in fact have the combined inclusion:

$$B \subseteq C[B] \subseteq B[C].$$

9.1.2 Preliminary results

Lemma 9.3. (see Shawyer [250], lemma 1). *There is a positive number θ, such that, as $x \to \infty$,*

$$e^{-x}\{u_{\alpha,\beta}(x) - u_{\alpha,\beta-1}(x)\} = \mathbf{O}(e^{-\theta x}).$$

This result is easily verified using the procedure in Hardy's book **T**[17], pp. 197–198], and the method of lemma 3 (c) in Borwein [37]. Note that Borwein 's lemma is only stated for $\alpha > 4$, because a specific value for θ is derived. For our purposes, it is sufficient to know that $\theta > 0$.

Lemma 9.4.

$$\sum_{n=N}^{\infty} \frac{n^{-k}x^n}{\Gamma(\alpha n + \beta)} \sim \alpha^{k-1}x^{(1-\beta-k)/\alpha}e^{x^{1/\alpha}}$$

as $x \to \infty$.

This is a special case of a more general result given in Borwein [40].

Lemma 9.5. (see Shawyer [250], lemma 3). *Suppose that r is a positive integer and that $\lambda_\nu = 0$ for $\nu = 0, -1, -2, \ldots$. If*

$$\mu_n = 1 + \sum_{m=1}^{r} \mu_{n,m}n^{-m} + \mathbf{O}\left(n^{-r-1}\right),$$

then there are constants $\lambda_\nu = \lambda_{\nu,n}$ ($\nu = 1, 2, \ldots, r$) such that for each nonnegative integer s,

$$\mu_n = \mu_n^{(s)} = 1 + \sum_{m=1}^{r} \frac{\Gamma(\alpha n + 1 - s)}{\Gamma(\alpha n + m + 1 - s}$$

$$\times \sum_{\nu=0}^{s}(-1)^{\nu}\binom{s}{\nu}\frac{(m+1)!}{(m-\nu)!}\lambda_{m-\nu}+\mathbf{O}\left(n^{-r-1}\right).$$

This result is readily verified.

From now on, suppose that $k > 0$. Let

$$\phi(x) = \alpha^{k-1}\sum_{n=0}^{\infty}\frac{\Gamma(n+k)x^{\alpha n}}{\Gamma(\alpha n+k)n!},$$

see (7.4). Then

$$\phi(x) = \sum_{n=0}^{\infty}\frac{\mu_n x^{\alpha n}}{\Gamma(\alpha n+1)},$$

where

$$\mu_n = \alpha^{k-1}\frac{\Gamma(n+k)\Gamma(\alpha n+1)}{\Gamma(n+1)\Gamma(\alpha n+k)}.$$

By Stirling's theorem, it follows that, for each positive integer r, there are constants $\mu_{n,m}$ such that

$$\mu_n = 1+\sum_{m=1}^{r}\mu_{n,m}n^{-m}+\mathbf{O}(n^{-r-1}),$$

and so μ_n satisfies the hypothesis of lemma 9.5 above.

Lemma 9.6. (see Shawyer [250], lemma 4). *For each positive integer r, there is a positive number θ, such that, as $t \to \infty$,*

$$D^r\left\{e^{-t}\phi(t)\right\} = \mathbf{O}\left(e^{-\theta t}\right)+\mathbf{O}\left(t^{-r-1}\right).$$

Proof. By Leibnitz's theorem,

$$\begin{aligned} D^r\{e^{-t}\phi(t)\} &= (-1)^r e^{-t}\sum_{s=0}^{r}\binom{r}{s}(-1)^s D^s\phi(t) \\ &= (-1)^r e^{-t}\sum_{s=0}^{r}\binom{r}{s}(-1)^s D^s\sum_{n=0}^{\infty}\mu_n^{(s)}\frac{t^{\alpha n}}{\Gamma(\alpha n+1)}. \end{aligned}$$

From lemma 9.5, and since

$$\sum_{s=\nu}^{r}\binom{r}{s}(-1)^s\binom{s}{\nu} = 0,$$

it follows that there are polynomials

$$p_{r,\nu}(t) = \sum_{m=0}^{\nu} p_{r,\nu,m} t^m$$

such that

$$\begin{aligned} D^r\{e^{-t}\phi(t)\} &= (-1)^r e^{-t} \sum_{\nu=-r+t}^{r} p_{r,\nu}(t)\{E_{\alpha,1-\nu}(t) - E_{\alpha,\nu}(t)\} \\ &\quad + \mathbf{O}\left(t^{-r-1}\right) \\ &= \mathbf{O}\left(e^{-\theta t}\right)\mathbf{O}\left(t^{-r-1}\right), \end{aligned}$$

where $\theta > 0$. Here, lemma 9.3 is used to give the first order term, and lemma 9.4 to give the second.

Lemma 9.7. (see Hardy and Riesz **T**[18], lemma 8). *Suppose that* $f(x) \to 0\,(C,k)$, *that* $0 < \delta < 1$, *and that* $k - \delta > 0$. *Then*

$$\int_0^u (x-t)^{\delta-1} f_{k-\delta}(t)dt = \mathbf{o}(u^k)$$

uniformly for $0 < u < x$.

9.1.3 Proof of the Cesàro product theorem

Assume, without loss of generality, that $s_n = 0$ for $n < N$ and that $\sigma = 0$. Thus $s_n \to 0\,(C,k)(B,\alpha,\beta)$; that is,

$$S_k(x) = \{\Gamma(k)\}^{-1} \int_0^x (x-1)^{k-1} S(t)dt = \mathbf{o}(x^k).$$

By lemma 7.4 and Borwein 's equivalence class of methods equivalent to the Borel-type method (B,α,β) (see (5.2)), it is sufficient to prove that

$$I(x) = \int_0^x h(x-t)S(t)dt = \mathbf{o}(x^k),$$

where $h(u) = u^{k-1}e^{-u}\phi(u)$.

Suppose first that k is a positive integer. Integration by parts k times yields that

$$I(x) = M \int_0^x S_k(t) D^k h(x-t)dt = \mathbf{o}(x^k) \int_\alpha^x |D^k h(t)|dt \quad (a > 0).$$

Now,

$$D^k h(t) = \sum_{r=0}^{k} \binom{k}{r} \{D^{k-r} t^{k-1}\}\{D^r e^{-t}\phi(t)\} = \sum_{r=1}^{k} M_{k,r} t^{r-1} D^r e^{-t}\phi(t),$$

and so it follows from lemma 9.6 that, as $t \to \infty$, $D^k h(t) = \mathbf{O}(e^{-\theta t}) + \mathbf{O}(t^{-2})$, where $\theta > 0$. Thus $D^k h(t) \in L[a, \infty)$, and so $I(x) = \mathbf{o}(x^k)$.

Suppose now that k is a positive nonintegral number. Let $[k]$ denote the greatest integer less than or equal to k, and let $\{k\} = k - [k] - 1$. In this case, integration by parts $[k]$ times yields that

$$\begin{aligned} I(x) &= M \int_0^x S_{[k]}(t) D^{[k]} h(x-t)dt \\ &= M \int_0^x \{t^{\{k\}} S_{[k]}(t)\}\{t^{-\{k\}} D^{[k]} h(t)\}dt \\ &= M\ S_k(x) + M \int_0^x Dt^{-\{k\}} D^{[k]} h(t)dt \int_0^t u_{\{k\}} S_{[k]}(x-u)du \\ &= \mathbf{o}(x^k) \left(M + \int_a^x |Dt^{-\{k\}} D^{[k]} h(t)|dt \right), \end{aligned}$$

by lemma 9.7, since $0 > \{k\} - 1$ and where $a > 0$. Further,

$$\begin{aligned} Dt^{-\{k\}} D^{[k]} t^{k-1} e^{-t}\phi(t) &= Dt^{-\{k\}} \sum_{r=0}^{[k]} \binom{[k]}{r} \{D^{[k]-r} t^{r-1}\}\{D^r e^{-t}\phi(t)\} \\ &= D \sum_{r=0}^{[k]} m_{k,r} t^r D^r e^{-t}\phi(t) \\ &= \sum_{r=0}^{[k]} M_{k,r} t^r D^{r+1} e^{-t}\phi(t), \end{aligned}$$

and so it follows from lemma 9.6 that for each $r = 0, 1, \ldots, [k]$ and as $t \to$

∞, $t^r D^{r+1} e^{-t}\phi(t) = \mathbf{O}(e^{-\theta t}) + \mathbf{O}(t^{-t})$, where $\theta > 0$. Thus $Dt^{-\{k\}}D^{[k]}h(t) \in L[a,\infty)$ and so $I(x) = \mathbf{o}(x^k)$.

The proof of the theorem is now complete.

9.2 Abelian relations with the Abel-type methods

As stated in Chapter 5, the Abel method is not included in the Borel method, and extra conditions are required for inclusion results. The same is true for the Abel-type methods and the Borel-type methods We give here the results developed by Shawyer and Yang [251] and [252].

9.2.1 Review of the definitions

We recall the notation of the binomial coefficients developed earlier:

$$E_n^\lambda = \binom{\lambda+n}{n} = \frac{(\lambda+1)(\lambda+2)\dots(\lambda+n)}{n!} \quad \text{for} \quad n = 0, 1, \dots,$$

$$E_0^\lambda = 1, \qquad E_n^\lambda = 0 \quad \text{for} \quad n = -1, -2, \dots .$$

We also recall that the Abel-type mean (for $y > 0$) is given by:

$$\sigma_\lambda(x) := (1+x)^{-\lambda-1} \sum_{n=0}^{\infty} E_n^\lambda\, s_n \left(\frac{x}{1+x}\right)^n .$$

This series is assumed to be convergent for all $x > 0$.

Definition 9.8. Ordinary Abel-type summability (see Borwein [34] or chapter 2 above). *If $\sigma_\lambda(x) \to \sigma$ as $x \to \infty$, then we say that the sequence $\{s_n\}$ is summable to σ by the Abel-type method of order λ, and write $s_n \to \sigma\ (A_\lambda)$.*

When $\lambda = 0$, we have the standard Abel summability method.

Definition 9.9. Strong Abel-type summability (see Mishra [192]). *If $p \geq 1$ and*

$$\int_0^x |\sigma_\lambda(t) - \sigma|^p \, dt = \mathbf{o}(x)$$

as $x \to \infty$, then we say that the sequence $\{s_n\}$ is strongly summable to σ by the Abel-type method of order λ and index p, and write $s_n \to \sigma\ [A_\lambda]_p$.

Definition 9.10. Absolute Abel-type summability (see Mishra [193]). *If $\sigma_\lambda(x) \to \sigma$ as $x \to \infty$ and if $\sigma_\lambda(x) \in BV_x[0,\infty)$, then we say that the sequence $\{s_n\}$ is summable to σ by the Abel-type method of order λ, and write $s_n \to \sigma\ |A_\lambda|$.*

9.2.2 Preliminary results

Lemma 9.11. (see Shawyer and Yang [251], lemma 1). *If*

$$F(w) = \int_a^b g(w,u)f(u)\,du \qquad (w \geq h > 0,\ \ -\infty < a < b < \infty),$$

and

$$\int_h^\infty |d_w g(w,u)| \leq g(u) \qquad (a < u < b),$$

then

$$\int_y^\infty |dF(w)| \leq \int_a^b g(u)|f(u)|\,du.$$

This is easily proved from first principles.

Lemma 9.12. (see Shawyer and Yang [251], lemma 2). *Suppose that m is a positive integer and that*

$$\mu_n = \prod_{i=1}^m \frac{b_i\,n + c_i}{d_i\,n + e_i} s_n \qquad (n = 0, 1, \ldots),$$

where b_i, c_i, d_i, and e_i are real with

$$\prod_{i=1}^m d_i \neq 0 \qquad \text{and} \qquad \prod_{i=1}^m (d_i\,n + e_i) \neq 0.$$

Then

$$\mu_n \to \prod_{i=1}^m \frac{b_i}{d_i} \sigma\ (A_\lambda) \text{ whenever } s_n \to \sigma\ (A_\lambda),$$

and

$$\mu_n \to \prod_{i=1}^{m} \frac{b_i}{d_i} \sigma \ |A_\lambda| \ \textit{whenever} \ s_n \to \sigma \ |A_\lambda| \ .$$

Proof. We give the proof for ordinary summability, the proof for absolute summability being similar.

Borwein [34] has shown that, if k is real, then $s_n/(n+k) \to 0 \ (A_\lambda)$ whenever $s_n \to \sigma \ (A_\lambda)$. Thus it follows that

$$\frac{bn+c}{dn+e} s_n = \frac{b}{d} s_n + \frac{cd-be}{d^2n+de} s_n \to \frac{b}{d} \sigma \ (A_\lambda) \ ,$$

for any real b, c, d, and e with $d \neq 0$. The result now follows immediately.

Lemma 9.13. (see Shawyer and Yang [251], lemmas 3 and 4). *Suppose that* $\sum_{n=0}^{\infty} a_n x^n$ *is convergent for* $|x| < 1$.

(i) *If* $s_n \to \sigma \, (B, \alpha, \beta)$, *then, for all* $\lambda > -1$,

$$\frac{\Gamma(\alpha n + \beta + \lambda)\Gamma(n+1)}{\Gamma(\alpha n + \beta)\Gamma(n+\lambda+1)} s_n \to \alpha^\lambda \sigma \ (A_\lambda) \ .$$

(ii) *If* $s_n \to \sigma \, |B, \alpha, \beta|$, *then, for all* $\lambda > -1$,

$$\frac{\Gamma(\alpha n + \beta + \lambda)\Gamma(n+1)}{\Gamma(\alpha n + \beta)\Gamma(n+\lambda+1)} s_n \to \alpha^\lambda \sigma \ |A_\lambda| \ .$$

Proof. First, we note that the convergence of $\sum_{n=0}^{\infty} a_n x^n$ for all $|x| < 1$ is sufficient to prove the convergence of the series defining $\sigma_\lambda(y)$ for all $y > 0$.
Part (i): let

$$\widehat{\sigma_\lambda}(y) = (1+y)^{-\lambda-1} \sum_{n=N}^{\infty} E_n^\lambda \frac{\Gamma(\alpha n + \beta + \lambda)\Gamma(n+1)}{\Gamma(\alpha n + \beta)\Gamma(n+\lambda+1)} s_n \left(\frac{y}{1+y}\right)^n ,$$

and

$$J(t) = \int_0^\infty e^{-u/t} u^\lambda S_{\alpha,\beta}(u) \, du \, , \qquad (0 < t < \infty) \, .$$

Now, if t and y are related by $\left(\frac{t}{1+t}\right)^{\alpha} = \frac{y}{1+y}$, we have

$$J(t) = \alpha\Gamma(\lambda+1)t^{\lambda+1}\left(\frac{t}{1+t}\right)^{\beta-1}\left(\frac{1+y}{1+t}\right)^{\lambda+1}\widehat{\sigma_\lambda}(y)\,.$$

Since $e^{-u/t}u^{\lambda}t^{-\lambda-1}/\Gamma(\lambda+1)$ is the kernel of a regular transformation, and since $S_{\alpha,\beta}(u) \to \sigma$ as $u \to \infty$, it follows that

$$\frac{J(t)}{\Gamma(\lambda+1)t^{\lambda+1}} \to \sigma$$

as $t \to \infty$. Further, $(1+y)/(1+t) \to 1/\alpha$ as $y \to \infty$, and since $t \to \infty$ if and only if $y \to \infty$, it follows that

$$\widehat{\sigma_\lambda}(y) \to \alpha^{\lambda}\sigma$$

as $y \to \infty$.

Part (ii): it remains only to show the bounded variation property. We note that

$$\widehat{\sigma_\lambda}(y) = \frac{1}{\alpha\Gamma(\lambda+1)}A_1(t)\,A_2(t)\,A_3(t)\,,$$

where

$$A_1(t) = \frac{J(t)}{t^{\lambda+1}}, \qquad A_2(t) = \left(\frac{1+t}{t}\right)^{\beta-1},$$

and

$$a_3(t) = \left(\frac{1-\{t/(1+t)\}^{\alpha}}{1-t/(1+t)}\right)^{\lambda+1}.$$

For $h > 0$, it is easy to see that $A_2(t)$ and $A_3(t)$ are of bounded variation with respect to t in $[h, \infty)$, whereas we use the given absolute summability and use lemma 9.11 to deduce the same for $A_1(t)$. This enables us to complete the proof of the lemma.

9.2.3 Theorems from Borel to Abel

All the theorems in this section assume that $\lambda > -1$ and that $\sum_{n=0}^{\infty} a_n x^n$ is convergent for $|x| < 1$.

Theorem 9.14. (see Shawyer and Yang [251], theorem 1). *If $s_n \to \sigma\ (B, \alpha, \beta)$ then $s_n \to \sigma\ (A_\lambda)$.*

Proof. Borwein [34] has shown that $s_n \to \sigma\ (A_\mu)$ whenever $s_n \to \sigma\ (A_\lambda)$ for all $\lambda > \mu > -1$. Thus, we assume without loss of generality, that λ is an integer. Thus, by lemma 9.13 (i), we have that

$$\frac{(\alpha n+\beta+\lambda-1)(\alpha n+\beta+\lambda-2)\dots(\alpha n+\beta+1)}{(n+\lambda)(n+\lambda-1)\dots(n+1)} s_n \to \alpha^\lambda \sigma\ (A_\lambda)\ .$$

Therefore, by lemma 9.12, we deduce that $s_n \to \sigma\ (A_\lambda)$ for all $\lambda > -1$.

Theorem 9.15. (see Shawyer and Yang [251], theorem 2). *If $s_n \to \sigma\,(B,\alpha,\beta)$ then $s_n \to \sigma\ [A_\lambda]_p$ for all $p \geq 1$.*

Proof. Mishra [192] has shown that $s_n \to \sigma\ [A_\lambda]_p$ if and only if $s_n \to \sigma\ (A_\lambda)$ and

$$\int_0^x \left| y\frac{d}{dy}\sigma_\lambda(y)\right|^p dy = \mathbf{o}(x)$$

as $x \to \infty$. Using theorem 9.14, we see that

$$y\frac{d}{dy}\sigma_\lambda(y) = (\lambda+1)\,(\sigma_{\lambda+1}-\sigma_\lambda) = \mathbf{o}(1)$$

as $y \to \infty$, and this is sufficient to complete the proof.

Theorem 9.16. (see Shawyer and Yang [251], theorem 3). *If $s_n \to \sigma\,[B,\alpha,\beta]_q$ then $s_n \to \sigma\ [A_\lambda]_p$ for all $q \geq 1$ and $p \geq 1$.*

This follows at once from the version of theorem 5.16 for $[B,\alpha,\beta]_q$ summability and from theorem 9.15.

Theorem 9.17. (see Shawyer and Yang [251]. theorem 4). *If $s_n \to \sigma\,|B,\alpha,\beta|$ then $s_n \to \sigma\ |A_\lambda|$.*

Proof. Mishra [193] has shown that $s_n \to \sigma\ |A_\mu|$ whenever $s_n \to \sigma\ |A_\lambda|$ for all $\lambda > \mu > -1$. Thus, we assume without loss of generality that λ is an integer. Thus, we may adapt the proof of theorem 9.14 using lemma 9.13 (ii) instead of lemma 9.13 (i).

9.3 Tauberian relations with the Abel-type methods

In this section, we shall give the results of Shawyer and Yang [252] that develop the conditions for deducing Borel-type summability from Abel-type summability.

Since Abel-type summability is a condition of the main results, it will be required to assume the convergence, for all $x \in (0,1)$, of $\sum_{n=0}^{\infty} E_n^{\lambda} s_n x^n$. Thus it follows that $s_n = \mathbf{O}((1+\delta)^n)$ for each $\delta > 0$. From this latter condition, it is easy to deduce that the series defining $s_{\alpha,\beta}(x)$ is convergent for all $x > 0$.

9.3.1 Preliminary results

Lemma 9.18. (see Shawyer and Yang [252], lemma 4). *Suppose that, for* $n = N, N+1, \ldots,$

$$\mu_n = \frac{\Gamma(\alpha n + \beta + \lambda)\Gamma(n+1)}{\Gamma(\alpha n + \beta)\Gamma(n + \lambda + 1)}.$$

Then $s_n \to \sigma\ (A_\lambda)$ *if and only if* $\mu_n s_n \to \alpha^\lambda \sigma\ (A_\lambda)$ *where* $\lambda > -1$.

The proof of this technical lemma is omitted. Details may be found in Shawyer and Yang [252]. The proof requires properties of Borwein 's $\overline{m}$-sequences and power series methods.

Lemma 9.19. (Pitt's form of Wiener's theorem: see, for example, Hardy **T** [17], theorem 233).
Suppose that $G(u)$ *is integrable over* $(0,\infty)$, *that*

$$\int_0^\infty u^{-ix} G(u)\, du \neq 0$$

for any real x, *that* $F(u)$ *is bounded and slowly decreasing with respect to* $\log(x)$, *and that*

$$\frac{1}{t}\int_0^\infty G\left(\frac{u}{t}\right) F(u)\, du \to \sigma \int_0^\infty G(u)\, du$$

as $t \to \infty$. *Then* $F(u) \to \sigma$ *as* $u \to \infty$.

9.3.2 Theorems from Abel to Borel

Clearly, there is no loss in assuming that a_n is real. Therefore we shall assume this in this section.

The first result in this direction was given by Doetsch [75].

Theorem 9.20. *If* $s_n \to \sigma\,(A)$ *and*

$$xe^{-x}\sum_{n=0}^{\infty}\frac{a_{n+1}x^n}{n!} \geq -M$$

where M *is a constant, then* $s_n \to \sigma\,(B)$.

The next theorem, theorem 1 in Shawyer and Yang [252], not only includes Doetsch's result as a special case, but gives a definite improvement.

Theorem 9.21. *If* $\lambda > -1$, $s_n \to \sigma\,(A_\lambda)$, *and* $S_{\alpha,\beta}(x)$ *is slowly decreasing with respect to* $\log(x)$, *then* $s_n \to \sigma\,(B,\alpha,\beta)$.

Proof. The proof of this theorem is quite complex. The essential difficulty lies in showing that $S_{\alpha,\beta}(x)$ is bounded, because we wish to make use of lemma 9.19. If we were to restrict consideration to nonnegative values of λ alone, the proof could be much simpler. For we could then use theorem 105 in Hardy **T**[17] instead of lemma 9.19. However, the proof that we give here is valid for all $\lambda > -1$.

We shall first show that the conditions of the theorem imply that $S_{\alpha,\beta}(x)$ is bounded for all $x \geq 0$ (compare theorem 10 in Pitt **T**[22]).

In order to simplify the notation, we shall, in this theorem, denote $S_{\alpha,\beta}\,(e^t)$ by $f_{\alpha,\beta}(t)$, $-\infty < t < \infty$. Since $v > t \to \infty$ and $0 \leq v - t \to 0$ if and only if $e^v > e^t \to \infty$ and $e^{v-t} = e^v/e^t \to 1$, it follows that $f_{\alpha,\beta}(t)$ satisfies the conditions of lemma 5.39, and so that there are positive numbers M_1, M_2, and t_0 such that, for all $t \geq t_0$,

$$f_{\alpha,\beta}(v) - f_{\alpha,\beta}(t) \geq -M_1(v-t) - M_2\,. \tag{9.1}$$

Set $k(v) = \dfrac{e^{v(\lambda+1)}e^{-e^v}}{\Gamma(\lambda+1)}$ for $(-\infty < v < \infty)$ so that $\displaystyle\int_{-\infty}^{\infty} k(v)\,dv = 1$.
Since, for given $\epsilon > 0$, there exist positive numbers t_1 and δ such that

$$f_{\alpha,\beta}(v) - f_{\alpha,\beta}(t) \geq -\epsilon \text{ for } t \geq t_1 \text{ and } 0 \leq v - t < \delta\,,$$

and

$$f_{\alpha,\beta}(t) - f_{\alpha,\beta}(v) \geq -\epsilon \text{ for } t \geq t_1 \text{ and } 0 \leq t - v < \delta\,,$$

we can choose an integer p such that

$$\frac{1}{2} < \int_{-\delta p}^{\delta p} k(v)\,dv = \gamma\,,$$

say, and so obtain that

$$f_{\alpha,\beta}(v) - f_{\alpha,\beta}(t) \geq -p\epsilon \text{ for } t \geq t_1 \text{ and } 0 \leq v - t < \delta p,$$

and

$$f_{\alpha,\beta}(t) - f_{\alpha,\beta}(v) \geq -p\epsilon \text{ for } t \geq t_1 \text{ and } 0 \leq t - v < \delta p.$$

Set $g(t) = \int_{-\infty}^{\infty} k(v-t) f_{\alpha,\beta}(v)\, dv$ for $-\infty < t < \infty$, and set $u = e^t$, so that

$$g(t) = \alpha \left(\frac{u}{1+u}\right)^{\beta-1} \left(\frac{1+y}{1+u}\right)^{\lambda+1} \sigma_\lambda^*(y),$$

where $\left(\frac{u}{1+u}\right)^{\alpha} = \frac{y}{1+y}$ and

$$\sigma_\lambda^*(y) = (1+y)^{-\lambda-1} \sum_{n=N}^{\infty} E_n^\lambda \frac{\Gamma(\alpha n + \beta + \lambda)\Gamma(n+1)}{\Gamma(\alpha n + \beta)\Gamma(n+\lambda+1)} s_n \left(\frac{y}{1+y}\right)^n.$$

By lemma 9.18, we obtain that $g(t) \to \sigma$ and $y \to \infty$. Since we may clearly choose N so that $\alpha N + \beta - 1 > 0$, we have that $g(t) \to 0$ as $t \to -\infty$. Thus we obtain that $g(t)$ is bounded for all t.

Now set $w(t) = \inf_{v \leq t} f_{\alpha,\beta}(v)$ and $W(t) = \sup_{v \leq t} f_{\alpha,\beta}(v)$. It follows from (9.1) for $v \geq t \geq t_0$, that $w(v) - w(t) \geq M_1(v-t) - M_2$.

In the rest of this proof, M denotes a constant, independent of r, s, t, and is not necessarily the same at each occurrence.

Now let $T = \max\{t_0, t_1\}$, and suppose that $t \geq T + \delta p$. Thus we have that

$$\begin{aligned} g(t) &= \left(\int_{t-\delta p}^{t+\delta p} + \int_{-\infty}^{t-\delta p} + \int_{t+\delta p}^{\infty}\right) k(v-t) f_{\alpha,\beta}(v)\, dv \\ &\geq \gamma \{f_{\alpha,\beta}(t-\delta p) - p\epsilon\} + w(t) \int_{-\infty}^{t-\delta p} k(v-t)\, dv \\ &\quad + f_{\alpha,\beta}(t-\delta p) \int_{t+\delta p}^{\infty} k(v-t)\, dv \\ &\quad + \int_{t+\delta p}^{\infty} k(v-t) \{f_{\alpha,\beta}(v) - f_{\alpha,\beta}(t-\delta p)\}\, dv \end{aligned}$$

$$\geq \gamma f_{\alpha,\beta}(t-\delta p) + w(t)\left(1 - \int_{t-\delta p}^{t+\delta p} k(v-t)\,dv\right)$$

$$-M_1 \int_{t+\delta p}^{\infty} (v-t)k(v-t)\,dv - M$$

$$\geq \gamma f_{\alpha,\beta}(t-\delta p) + (1-\gamma)w(t) - M\,,$$

since

$$0 < \int_{t+\delta p}^{\infty} (v-t)k(v-t)\,dv = \int_{\delta p}^{\infty} vk(v)\,dv$$

$$\leq \frac{1}{\Gamma(\lambda+1)} \int_1^{\infty} e^{-v}v^{\lambda}\log(v)\,dv < \lambda + 1\,,$$

so that $f_{\alpha,\beta}(t) + \left(\frac{1-\gamma}{\gamma}\right) w(t) \leq M$. Since $w(t)$ decreases, it follows that $W(t) + \left(\frac{1-\gamma}{\gamma}\right) w(t) \leq M$.

Further, we have that

$$g(t) = \left(\int_{t-\delta p}^{t+\delta p} + \int_{-\infty}^{t-\delta p} + \int_{t+\delta p}^{\infty}\right) k(v-t) f_{\alpha,\beta}(v)\,dv$$

$$\leq \{f_{\alpha,\beta}(t+\delta p) + p\epsilon\} \int_{t-\delta p}^{t+\delta p} k(v-t)\,dv + M \int_{t+\delta p}^{\infty} k(v-t)\,dv$$

$$+W(t+\delta p) \int_{-\infty}^{t+\delta p} k(v-t)\,dv$$

$$+\left(\frac{1-\gamma}{\gamma}\right) \int_{t+\delta p}^{\infty} k(v-t)w(v)\,dv$$

$$\leq \gamma f_{\alpha,\beta}(t+\delta p) - \left(\frac{1-\gamma}{\gamma}\right) w(t+\delta p) \left(\int_{-\infty}^{t-\delta p} + \int_{t+\delta p}^{\infty}\right) k(v-t)\,dv$$

$$-\left(\frac{1-\gamma}{\gamma}\right) \int_{t+\delta p}^{\infty} k(v-t)\{w(v) - w(t)\}\,dv + M$$

$$\leq \gamma f_{\alpha,\beta}(t+\delta p) - \left(\frac{(1-\gamma)^2}{\gamma}\right) w(t+\delta p) + M\,,$$

so that $\left(\frac{(1-\gamma)}{\gamma}\right)^2 w(t) + M \leq f_{\alpha,\beta}(t)$.

Now, if $w(t) \to -\infty$ as $t \to \infty$, then there exists a sequence $\{t_n\}$ such that $f_{\alpha,\beta}(t_n) = w(t_n) \to -\infty$. Since $f_{\alpha,\beta}(t) \geq M + \left(\frac{1-\gamma}{\gamma}\right)^2 w(t)$, we have that $\left(1 - \left(\frac{1-\gamma}{\gamma}\right)^2\right) w(t_n) \geq M$, which is a contradiction, since M is a constant and $\gamma > 1/2$. Therefore, $w(t)$ is bounded below. It now follows from the inequality:

$$M + \left(\frac{(1-\gamma)}{\gamma}\right)^2 w(t) \leq f_{\alpha,\beta}(t) \leq M - \left(\frac{(1-\gamma)}{\gamma}\right) w(t),$$

that $f(t)$ is bounded for all $t \geq T + \delta p$.

Further, since $f_{\alpha,\beta}(x)$ is clearly bounded for all $x < T + \delta p$, we have that $f_{\alpha,\beta}(x)$ is bounded for all x. Since $S_{\alpha,\beta}(x) = f_{\alpha,\beta}(\log(x))$, we deduce that $S_{\alpha,\beta}(x)$ is bounded for all $x > 0$. This completes the first part of the proof.

Now, let $G(u) = \dfrac{e^{-u}u^{\lambda}}{\Gamma(\lambda+1)}$ for $u > 0$, so that

$$g(t) = \frac{1}{t}\int_0^{\infty} g\left(\frac{u}{t}\right) S_{\alpha,\beta}(u)\,du$$

and

$$\int_0^{\infty} G(u)\,du = \frac{1}{\Gamma(\lambda+1)}\int_0^{\infty} e^{-u}u^{\lambda}\,du = 1\,.$$

Further, $\int_0^{\infty} G(u)u^{-ix}\,du \neq 0$ for any real x.

Thus we have satisfied all the hypotheses of lemma 9.19 with $S_{\alpha,\beta}(x)$ in place of $F(x)$, so that we can conclude that $S_{\alpha,\beta}(x) \to \sigma$ as $x \to \infty$, that is, that $s_n \to \sigma\,(B,\alpha,\beta)$.

Theorem 9.22. (see Shawyer and Yang [252], theorem 2). *If $\lambda > -1$, $s_n \to \sigma\ (A_\lambda)$ and $xS'_{\alpha,\beta}(x) \geq -M$ for $x \geq h \geq 0$, where M is a constant, then $s_n \to \sigma\,[B,\alpha,\beta]$.*

Proof. First, as noted above, the conditions of the theorem imply that $S_{\alpha,\beta}(x)$ is slowly decreasing with respect to $\log(x)$. We therefore have, from theorem 9.21, that $s_n \to \sigma\,(B,\alpha,\beta)$.

From the condition $xS'_{\alpha,\beta}(x) \geq -M$, it is easy to obtain that $|S'_{\alpha,\beta}(x)| \leq S'_{\alpha,\beta}(x) + M/x$ for $x \geq h$. Thus

$$\int_0^x e^t \left|S'_{\alpha,\beta}(t)\right| dt \leq M \int_h^x \frac{e^t}{t}\, dt + \int_h^x e^t \left|S'_{\alpha,\beta}(t)\right| dt$$
$$+ \int_0^h e^t \left|S'_{\alpha,\beta}(t)\right| dt$$
$$= I_1 + I_2 + I_3 ,$$

say, where we easily see that both $I_2 = \mathbf{o}(e^x)$ and $I_3 = \mathbf{o}(e^x)$. Also,

$$|I_1| = \left|e^x S_{\alpha,\beta}(x) - e^h S_{\alpha,\beta}(h) - \int_h^x e^t S_{\alpha,\beta}(t)\, dt\right|$$
$$\leq e^x \left|S_{\alpha,\beta}(x) - \sigma\right| + e^h \left|S_{\alpha,\beta}(h)\right| + \int_h^x e^t \left|S_{\alpha,\beta}(x) - \sigma\right| dt$$
$$= \mathbf{o}(e^x) .$$

It now follows from theorem 5.21 that $s_n \to \sigma\,[B, \alpha, \beta]$.
Note. The condition $xS'_{\alpha,\beta}(x) \geq -M$ may, of course, be replaced by the condition $xS'_{\alpha,\beta}(x) \leq M$.

Theorem 9.23. (see Shawyer and Yang [252], theorem 2). *If $\lambda > -1$, $s_n \to \sigma\ (A_\lambda)$ and there exists a function $\phi(x)$, positive and integrable over $[h, \infty)$ such that $S'_{\alpha,\beta}(x) \geq -M\phi(x)$ and $x\phi(x) \geq M$ for $x \geq h \geq 0$, where M is a constant, then $s_n \to \sigma\,|B, \alpha, \beta|$.*

Proof. First, as noted above, the conditions of the theorem imply that $xS'_{\alpha,\beta}(x) > -M$, and so it follows that $s_n \to \sigma\,(B, \alpha, \beta)$.

It is easy to obtain that $|S'_{\alpha,\beta}(x)| \leq S'_{\alpha,\beta}(x) + M\phi(x)$ for $x \geq h$, and so it follows that

$$\int_h^\infty |S'_{\alpha,\beta}(x)|\, dx \leq \int_h^\infty S'(x)\, dx + M \int_h^\infty \phi(x)\, dx .$$

Also, since $S'_{\alpha,\beta}(x)$ is continuous on $[0, h]$, we obtain that $S(x) \in BV_x[0, \infty)$, yielding that $s_n \to \sigma\,|B, \alpha, \beta|$.
Note. The conditions of the theorem may be modified to permit two inequalities in the reverse direction.

Further, note that there is a natural hierarchy in the three theorems in this section: the conditions of theorem 9.23 imply those of theorem 9.22, which in turn, imply those of theorem 9.21. The same can be said for the conclusions of the three theorems.

9.4 Tauberian relations with the logarithmic method

In this section, we shall give the results of Borwein and Watson [56] that develop the conditions for deducing Borel-type summability from logarithmic summability. Since the logarithmic method is included in all the Abel-type methods, it is not surprising that the tauberian condition needed is more delicate.

We begin by reviewing the definition of the logarithmic method of summability:

Definition 9.24. *If* $L(x) := \dfrac{-1}{\log(1-x)} \sum_{n=0}^{\infty} \dfrac{s_n}{n+1} x^{n+1}$ *exists for* $|x| < 1$, *and if* $\lim_{x\to 1^-} L(x) = \sigma$, *then we say that the sequence* $\{s_n\}$ *is summable by the logarithmic method to* σ, *and write* $s_n \to \sigma\,(L)$.

This regular method, introduced by Borwein [36], satisfies the following properties:

Theorem 9.25. (see Borwein [36]). *If* $s_n \to \sigma\ (A_\lambda)$, *then* $s_n \to \sigma\,(L)$.

Theorem 9.26. (see Borwein and Watson [57]). *If* $s_n \to \sigma\,(L)$ *and* $\sigma_\lambda(x)$ *is slowly decreasing with respect to* $\log\log(x)$, *then* $s_n \to \sigma\ (A_\lambda)$.

Note that this slowly decreasing condition is more delicate than any other slowly decreasing condition encountered so far.

9.4.1 Preliminary results

Here we shall give two technical lemmas and one technical theorem. The lemmas can be proved by straightforward calculations. The theorem is a functional version of Vijayaraghavan's theorem. See [301].

For notational ease, we define, for $t \geq a > 1$,

$$J(t) := \frac{1}{\log(t)} \int_a^{\infty} \frac{e^{-u/t}}{u} S_{\alpha,\beta}(u)\, du\,.$$

Lemma 9.27. (see Borwein and Watson [56], lemma 2).

(i) *If* $s_n \to \sigma\,(L)$, *then* $J(t) \to \sigma$ *as* $t \to \infty$,

(ii) *If* $L(x)$ *exists for* $|x| < 1$ *and* $J(t) \to \sigma$ *as* $t \to \infty$, *then* $s_n \to \sigma\,(L)$.

Lemma 9.28. (see Borwein and Watson [56], lemma 3). *Suppose that* $\gamma > 1, t > 1, a > 0$. *Then*

(i) $$\frac{1}{\log(t)} \int_t^\infty \frac{e^{-u/t}}{u}\, du \to 0 \quad as \quad t \to \infty\,,$$

(ii) $$\frac{1}{\log(t)} \int_a^t \frac{e^{-u/t}}{u}\, du \to 1 \quad as \quad t \to \infty\,,$$

(iii) $$0 < \int_a^t \left(e^{-u/t^\gamma} - e^{-u/t}\right) \frac{du}{u} < 1\,,$$

(iv) $$\frac{1}{\log(t)} \int_t^{t^\gamma} \frac{e^{-u/t^\gamma}}{u}\, du \to \gamma - 1 \quad as \quad t \to \infty\,.$$

Theorem 9.29. (see Borwein and Watson [55]; compare Vijayaraghavan [301], which is also to be found in Hardy **T**[17]). *Suppose that the following conditions hold:*

(1) $K(t,u)$ *is defined, real-valued and nonnegative for* $t > a, u \geq a$; *moreover* $\int_a^\infty K(t,u)\, du$ *exists as a Lebesgue integrals for each* $t > a$,

(2) $\int_a^\infty K(t,u)\, du \to 1$ *as* $t \to \infty$,

(3) f *is real-valued and continuous on* $[a, \infty)$,

(4) $F(t) := \int_a^\infty K(t,u) f(u)\, du$ *exists as a Cauchy–Lebesgue integral for each* $t > a$,

(5) f *is slowly decreasing with respect to* ϕ,

(6) $\phi(t) - \phi(t-1) \to 0$ *as* $t \to \infty$,

(7) $\int_a^x K(t,u)\, du \to 0$ *whenever* $t \geq x \to \infty$ *and* $\phi(t) - \phi(x) \to \infty$,

(8) $\int_x^\infty K(t,u)\,(\phi(u) - \phi(x))\, du \to 0$ *whenever* $x \geq t \to \infty$ *and* $\phi(x) - \phi(t) \to \infty$, *and*

(9) $F(t) = \mathbf{O}(1)$ *for* $t > a$.

Then $f(u) = \mathbf{O}(1)$ *for* $u > a$.

9.4.2 The logarithmic theorem

Theorem 9.30. (see Borwein and Watson [56], theorem 1). *If* $s_n \to \sigma\,(L)$ *and* $S_{\alpha,\beta}(x)$ *is slowly decreasing with respect to* $\log\log(x)$, *then* $s_n \to \sigma\,(B, \alpha, \beta)$.

Proof. In order to use theorem 9.29, we set

$$a = 1 + e^e\,,$$

$$K(t,u) = \begin{cases} \dfrac{1}{\log(t)}\,\dfrac{e^{-u/t}}{u} & \text{for} \quad t \geq a\,,\, u \geq a\,, \\ 0 & \text{otherwise,} \end{cases}$$

$$\phi(t) = \log\log(t) \quad \text{for} \quad t \geq a\,,$$

$$f(u) = S_{\alpha,\beta}(u) \quad \text{for} \quad u \geq a\,.$$

Then

$$\int_a^\infty K(t,u) f(u)\, du = J(t) \quad \text{for} \quad t \geq a\,.$$

We first show that the conditions of the theorem imply that $S_{\alpha,\beta}(x) = \mathbf{O}(1)$ for $u > a$.

Conditions (1), (3), (5), and (6) of theorem 9.29 clearly hold. Parts (i) and (ii) of lemma 9.28 imply that condition (2) holds. Further, the logarithmic summability of $\{s_n\}$ implies that conditions (4) and (9) hold. It therefore remains to show that conditions (7) and (8) hold.

We observe that

$$\int_a^\infty K(t,u)\, du \leq \frac{1}{\log(t)} \int_a^x \frac{du}{u} = \frac{\log(x) - \log(a)}{\log(t)} \to 0$$

as $t \geq x \to \infty$ and $\log\log(t) - \log\log(x) \to \infty$. This shows that (7) holds. Further, we note that

$$\begin{aligned} &\int_x^\infty K(t,u)\,(\phi(u) - \phi(x))\, du \\ &= \frac{1}{\log(t)} \int_x^\infty \frac{e^{-u/t}}{u}\,(\log\log(u) - \log\log(x))\, du \\ &\leq \frac{1}{\log(t)} \int_x^\infty \frac{e^{-u/t}}{u} \left(\frac{u - x}{x \log(x)}\right) du \end{aligned}$$

$$
\begin{aligned}
&\leq \frac{1}{x\log(x)\log(t)}\int_t^\infty e^{-u/t}\,du \\
&= \frac{te^{-x/t}}{x\log(x)\log(t)} \to 0 \quad \text{as} \quad x \geq t \to \infty\,,
\end{aligned}
$$

so that (8) holds. This means that we have proved that $S_{\alpha,\beta}(u)$ is bounded in (a, ∞).

Without loss of generality, we suppose that $\sigma = 0$. Therefore, by lemma 9.27, $J(t) \to 0$ as $t \to \infty$. All that remains is to show that $S_{\alpha,\beta}(u) \to 0$ as $u \to \infty$.

Choose $\epsilon > 0$. Since $S_{\alpha,\beta}(u)$ is slowly decreasing with respect to $\phi(x) = \log\log(x)$, there exist numbers $x \geq a$ and $\delta > 0$ such that $S_{\alpha,\beta}(u) - S_{\alpha,\beta}(t) > -\epsilon$ whenever $u > t \geq x$ and $\log\log(u) - \log\log(t) < \delta$. Equivalently, setting $\gamma = e^\delta$, we have

$$
S_{\alpha,\beta}(t) - \epsilon < S_{\alpha,\beta}(u) \quad \text{whenever} \quad x < t < u < t^\gamma\,. \tag{9.2}
$$

This equation implies that, for $t > x$,

$$
\begin{aligned}
I_1 &:= \frac{1}{\log(t)}\int_t^{t^\gamma} \frac{e^{-u/t^\gamma}}{u}\left(S_{\alpha,\beta}(t) - \epsilon\right)\,du \\
&\leq \frac{1}{\log(t)}\int_t^{t^\gamma} \frac{e^{-u/t^\gamma}}{u} S_{\alpha,\beta}(u)\,du := I_2\,.
\end{aligned}
$$

Now, by lemma 9.28 and the established property that $S_{\alpha,\beta}(u) = \mathbf{O}(1)$, we obtain that

$$
\begin{aligned}
I_2 &= \gamma\, J\left(t^\gamma\right) - J(t) - \frac{\gamma}{\log\left(t^\gamma\right)}\int_{t^\gamma}^\infty \frac{e^{-u/t^\gamma}}{u} S_{\alpha,\beta}(u)\,du \\
&\quad - \frac{1}{\log(t)}\int_a^t \frac{e^{-u/t^\gamma} - e^{-u/t}}{u} S_{\alpha,\beta}(u)\,du \\
&\quad + \frac{1}{\log(t)}\int_t^\infty \frac{e^{-u/t^\gamma}}{u} S_{\alpha,\beta}(u)\,du \\
&= \mathbf{o}(1) \quad \text{as} \quad t \to \infty\,.
\end{aligned}
$$

Further, using part (iv) of lemma 9.28, we have

$$
I_1 = \left(S_{\alpha,\beta}(t) - \epsilon\right)\left(\gamma - 1 + \mathbf{o}(1)\right)\,.
$$

Therefore,

$$S_{\alpha,\beta}(t) - \epsilon \leq \frac{I_1}{\gamma - 1 + \mathbf{o}(1)} = \mathbf{o}(1)\,,$$

so that

$$\limsup_{t\to\infty} S_{\alpha,\beta}(t) \leq \epsilon\,. \tag{9.3}$$

Rewriting equation (9.2) gives

$$S_{\alpha,\beta}(u) < S_{\alpha,\beta}(t) + \epsilon \quad \text{whenever} \quad x < t^{1/\gamma} < u < t\,.$$

This equation implies that, for $t^{1/\gamma} \geq x$,

$$\begin{aligned} I_3 &:= \frac{1}{\log(t)} \int_{t^{1/\gamma}}^{t} \frac{e^{-u/t}}{u} S_{\alpha,\beta}(u)\,du \\ &\leq \frac{1}{\log(t)} \int_{t^{1/\gamma}}^{t} \frac{e^{-u/t}}{u} \left(S_{\alpha,\beta}(t) + \epsilon\right)\,du := I_4\,. \end{aligned}$$

By lemma 9.28 with t replaced by $t^{1/\gamma}$, and the established property that $S_{\alpha,\beta}(u) = \mathbf{O}(1)$, we obtain that

$$\begin{aligned} I_3 &= J(t) - \frac{1}{\gamma} J\left(t^{1/\gamma}\right) - \frac{1}{\log(t)} \int_{t}^{\infty} \frac{e^{-u/t}}{u} S_{\alpha,\beta}(u)\,du \\ &\quad - \frac{1}{\log(t)} \int_{a}^{t^{1/\gamma}} \frac{e^{-u/t} - e^{-u/t^{1/\gamma}}}{u} S_{\alpha,\beta}(u)\,du \\ &\quad + \frac{1}{\log(t)} \int_{t^{1/\gamma}}^{\infty} \frac{e^{-u/t^{1/\gamma}}}{u} S_{\alpha,\beta}(u)\,du \\ &= \mathbf{o}(1) \quad \text{as} \quad t \to \infty\,. \end{aligned}$$

We also have that

$$I_4 = \left(S_{\alpha,\beta}(t) + \epsilon\right)\left(1 - \frac{1}{\gamma} + \mathbf{o}(1)\right)\,.$$

Therefore,

$$S_{\alpha,\beta}(t) + \epsilon \geq \frac{I_3}{1 - 1/\gamma + \mathbf{o}(1)} = \mathbf{o}(1)\,,$$

so that

$$\liminf_{t\to\infty} S_{\alpha,\beta}(t) \geq \epsilon\,.$$

It follows from this and (9.3) that $S_{\alpha,\beta}(t) \to 0$ as $t \to \infty$, and the proof is complete.

9.5 Relations with the Lambert method

In this section, we discuss the property that sequences which are summable both by Abel's method and by Borel's method, are also summable by Lambert's method.

We start this section by giving, in a convenient way, the definitions of the three summability methods involved. The notation given here for Abel's method is not the usual one, and is used only in this section. The definition of Lambert's method is not given earlier, and the definition of Borel's method (B) is repeated for completeness.

The Abel, Lambert, and Borel means are given by:

$$A(v) := \sum_{n=0}^{\infty} a_n v^{-n}\,, \tag{9.4}$$

$$L(v) := a_0 - (v-1)\sum_{n=1}^{\infty} A'\left(v^n\right) v^n\,, \tag{9.5}$$

$$B(v) := e^{-v}\sum_{n=0}^{\infty} \frac{s_n}{n!}\, v^n\,. \tag{9.6}$$

Definition 9.31.

(a) *If* (9.4) *is convergent for* $|v| > 1$ *and if* $\lim_{v\to1+} A(v) = \sigma$, *then we say that* $\sum_{n=0}^{\infty} a_n$ *(or* $\{s_n\}$*) is summable to* σ *by Abel's method, and write* $\sum_{n=0}^{\infty} a_n = \sigma\,(A)$ *(or* $s_n \to \sigma\,(A)$ *).*

(b) *If* (9.5) *is convergent for* $|v| > 1$ *and if* $\lim_{v\to1+} L(v) = \sigma$, *then we say that* $\sum_{n=0}^{\infty} a_n$ *(or* $\{s_n\}$*) is summable to* σ *by Lambert's method, and write* $\sum_{n=0}^{\infty} a_n = \sigma\,(L)$ *(or* $s_n \to \sigma\,(L)$ *).*

(c) *If* (9.6) *is convergent for all* $v \geq 0$ *and if* $\lim_{v \to \infty} S(v) = \sigma$, *then we say that* $\sum_{n=0}^{\infty} a_n$ (*or* $\{s_n\}$) *is summable to* σ *by Borel's method, and write* $\sum_{n=0}^{\infty} a_n = \sigma\,(B)$ (*or* $s_n \to \sigma\,(B)$).

Let $\mathcal{A}$ represent the set of all sequences that are summable (A). We use $\mathcal{L}$ and $\mathcal{B}$ with similar meanings.

Hardy and Littlewood [114] have shown the inclusion $\mathcal{L} \subset \mathcal{A}$. They deduced this relation from the convergence of $\sum_{n=1}^{\infty} \frac{|N(n)|}{n}$, where we have that $N(x) := \sum_{n \leq x} \frac{\mu(n)}{n}$ and that $\mu(n)$ is the Möbius function. Subsequently, Hoischen [121] has proved that this inclusion is equivalent to the convergence of $\sum_{n=1}^{\infty} \frac{|N(n)|}{n}$, and that this fact is somewhat deeper that the prime number theorem. Further, Hoischen [123] has proved the equivalence of the Riemann hypothesis with certain saturation properties between Lambert's method and Abel's method.

Since $\mathcal{B}$ contains series with the property $\limsup_{n \to \infty} |a_n|^{1/n} > 1$, we have that $\mathcal{B} \not\subseteq \mathcal{A}$ and $\mathcal{B} \not\subseteq \mathcal{L}$ since the series defining the Abel and Lambert means will not be convergent.

However, if $\limsup_{n \to \infty} |a_n|^{1/n} < \infty$, then $A(v)$ has an analytic continuation into $\Re(v) > 1$ and further, $\lim_{v \to 1+} A(v) = \sigma$. Therefore it makes sense to ask about the properties of sequences in $\mathcal{A} \cap \mathcal{B}$. In particular, do such sequences necessarily lie in $\mathcal{L}$? This question will be answered in the following theorem.

Theorem 9.32. (see Gawronski, Siebert, and Trautner [94]).

$$\mathcal{A} \cap \mathcal{B} \subseteq \mathcal{L}.$$

The proof of this theorem makes use of functional analysis. This is the only exception to the standard classical analytic proofs presented in this book.

The proof is presented in three parts: the first is a general framework, the second gives the transformation formulae used, and the third establishes the theorem. Since both Abel and Borel summability are assumed, the following conditions are assumed for the remainder of this proof:

$$|B(v)| \le M \quad \text{where } M \text{ is a constant}$$

$$\limsup_{n\to\infty} |a_n|^{1/n} \le 1$$

$$\sum_{n=0}^{\infty} a_n \in \mathcal{A} \cap \mathcal{B} \tag{9.7}$$

$$a_0 = a_1 = 0 \tag{9.8}$$

$$\lim_{v\to\infty} B(v) = 0 \tag{9.9}$$

Assumptions (9.8) and (9.9) may be made without loss of generality since all three summability methods are regular. We shall also need to use a weight function

$$w(x) := \min_{x>0} \left\{x^2, 1\right\},$$

and the Banach spaces

$$U_1 := \left\{ f(x) \mid f(x) \text{ is continuous for } x > 0,\ \| f \| = \sup_{x>0} \left| \frac{f(x)}{w(x)} \right| < \infty \right\},$$

$$U_2 := \left\{ g(v) \mid g(v) \text{is continuous for } v \in (1,2],\ \| g \| = \sup_{1<v\le 2} |g(v)| < \infty \right\}.$$

Note that if (9.7) and (9.8) hold, then $B(x) \in U_1$ and if $\{s_n\} \in \mathcal{L}$, then $L(v) \in U_2$.

We shall show the existence of a kernel $K_{L,B}(v,x)$ with the following property:

$$g(v) = \int_0^\infty K_{L,B}(v,x) f(x)\, dx\,, \qquad 1 < v \le 2\,, \tag{9.10}$$

defines a linear map from U_1 into the space of continuous functions on the interval $(1,2]$ satisfying

$$L(v) = \int_0^\infty K_{L,B}(v,x) B(x)\, dx\,. \tag{9.11}$$

It is critical to the proof to show that (9.10) is a bounded map from U_1 into U_2, that is (see lemma 9.36),

$$\| K_{L,B}(v,x) \| = \sup_{1<v\leq 2} \int_0^\infty |K_{L,B}(v,x)|\, w(x)\, dx < \infty . \tag{9.12}$$

Finally, it is shown that

$$\lim_{v\to 1^+} K_{L,B}(v,x) = 0 \tag{9.13}$$

uniformly on any interval $[x_1, x_2], 0 < x_1 < x_2 < \infty$. From (9.11), (9.12), and (9.13), using the standard techniques as in the Toeplitz–Kojima–Schur theorem, we show that (9.7), (9.8), and (9.9) imply that

$$\lim_{v\to 1^+} L(v) = 0 . \tag{9.14}$$

9.5.1 Transformation formulae

Here, we give the transformations relating the various means under consideration. It is assumed, of course, that $\{s_n\} \in \mathcal{A}$ and $\{s_n\} \in \mathcal{B}$.

Lemma 9.33. (see Gawronski, Siebert, and Trautner [94], lemma 1). *If x is real and $R > 1$, then*

$$B(x) = \oint_{|v|=R} K_{B,A}(x,v) A(v)\, dv , \tag{9.15}$$

with

$$K_{B,A}(x,v) = \frac{e^{-x(1-v)}}{2\pi i (v-1)} . \tag{9.16}$$

Proof. From (9.4), we have that

$$\frac{A(v)}{v-1} = \sum_{n=0}^{\infty} \frac{s_n}{v^{n+1}} ,$$

and further, that

$$s_n = \frac{1}{2\pi i} \oint_{|v|=R} \frac{A(v)}{v-1} v^n\, dv .$$

This implies that

$$B(x) = e^{-x} \sum_{n=0}^{\infty} \frac{s_n x^n}{n!} = e^{-x} \sum_{n=0}^{\infty} \frac{x^n}{n!} \frac{1}{2\pi i} \oint_{|v|=R} \frac{A(v)}{v-1} v^n \, dv$$

$$= \oint_{|v|=R} \frac{1}{2\pi I(v-1)} e^{-x} \sum_{n=0}^{\infty} \frac{(xv)^n}{n!} \, dv \, .$$

Lemma 9.34. (see Gawronski, Siebert, and Trautner [94], lemma 2). *For* $\Re(v) > 1$

$$A(v) = \int_0^{\infty} K_{A,B}(v,x) B(x) \, dx \tag{9.17}$$

with

$$K_{A,B}(v,x) = (v-1) e^{-(v-1)} \, . \tag{9.18}$$

Proof. First note that $a_0 = 0$, and observe that

$$\int_0^{\infty} K_{A,B}(v,x) K_{B,A}(x,t) \, dx = (v-1) \int_0^{\infty} e^{-x(v-1)} \frac{e^{-x(1-t)}}{2\pi i(t-1)} \, dx$$

$$= \frac{1}{2\pi i} \frac{(v-1)}{(t-1)(v-t)} \, ,$$

which is valid for $\Re(v-t) > 0$. Hence, for $v > R > 1$, from lemma 9.33, we obtain that

$$\int_0^{\infty} K_{A,B}(v,x) B(x) \, dx = \int_0^{\infty} K_{A,B}(v,x) \, dx \oint_{|t|=R} K_{B,A}(x,t) A(t) \, dt$$

$$= \frac{1}{2\pi i} \oint_{|t|=R} A(t) \frac{(v-1)}{(t-1)(v-t)} \, dt = A(v) \, .$$

Lemma 9.35. (see Gawronski, Siebert, and Trautner [94], lemma 3). *Assuming condition* (9.8), *we have for* $1 < v \leq 2$,

$$L(v) = \int_0^{\infty} K_{L,B}(v,x) B(x) \, dx \tag{9.19}$$

with

$$K_{L,B}(v,x) = -(v-1)\sum_{n=1}^{\infty} e^{-x(v^n-1)}\left(1 - x(v^n-1)\right)v^n . \tag{9.20}$$

Proof. If we differentiate (9.17), we obtain that

$$\begin{aligned} A'(v) &= \int_0^{\infty} \frac{\partial}{\partial v} K_{A,B}(v,x) B(x)\, dx \\ &= \int_0^{\infty} e^{-x(v-1)}\left(1 - x(v-1)\right) B(x)\, dx , \end{aligned}$$

and further, from (9.5), we obtain that

$$\begin{aligned} L(v) &= -(v-1)\sum_{n=1}^{\infty} A'(v^n)\, v^n \\ &= -(v-1)\sum_{n=1}^{\infty}\int_0^{\infty} e^{-x(v^n-1)}\left(1 - x(v^n-1)\right) v^n B(x)\, dx . \end{aligned}$$

This will prove the lemma provided that we can justify the exchange of sum and integral. To do this, we note that, for $x > 0$ and $1 < v \le 2$,

$$\begin{aligned} \frac{1}{v-1}\left|K_{L,B}(v,x)\right| &\le \sum_{n=1}^{\infty} e^{-x(v^n-1)}\left(1 + x(v^n-1)\right) v^n \\ &= \sum_{n=1}^{\infty} e^{-x(v^n-1)}\left(1 + (v^n-1)(x+1) + (v^n-1)^2 x\right) . \end{aligned}$$

We now use the fact that $y^m e^{-y} \le M e^{-y/2}, y > 0 (m = 0, 1, 2)$, and $(v^n - 1) \ge n(v-1)$ to see that, for $x > 0$ and $1 < v \le 2$,

$$\begin{aligned} \frac{1}{v-1}\left|K_{L,B}(v,x)\right| &\le M\left(1 + \frac{1+x}{x} + \frac{1}{x}\right)\sum_{n=1}^{\infty} e^{-x(v^n-1)/2} \\ &\le 2M\frac{1+x}{x}\,\frac{e^{-x(v-1)/2}}{1 - x^{-x(v-1)/2}} . \end{aligned}$$

Using $e^{-t}\left(1-e^{-t}\right)^{-1} \le e^{-t}(1+t)t^{-1}$ for $t>0$, we further obtain that, for $x>0$ and $1<v\le 2$,

$$|K_{L,B}(v,x)| \le 4M\left(\frac{1+x}{x}\right)^2 e^{-x(v-1)/2}\,. \tag{9.21}$$

In view of (9.8), there exists a constant M' such that

$$|B(x)| \le M'\left(\frac{x}{1+x}\right)^2$$

for $x>0$, which in turn implies that

$$\int_0^\infty |K_{L,B}(v,x)B(x)|\,dx \le 4MM'\int_0^\infty \left(\frac{1+x}{x}\right)^2\left(\frac{x}{1+x}\right)^2 e^{-x(v-1)/2}\,dx,$$

which is finite. Now it is clear that an application of Fubini's theorem justifies the interchange, as required.

9.5.2 Essential lemmas

First we show that (9.10) defines a bounded map from the Banach space U_1 into U_2. To do this we prove the following lemma:

Lemma 9.36. (see Gawronski, Siebert, and Trautner [94], lemma 4). *The norm of $K_{L,B}$ is finite.*

Proof. We must show that (9.12) is true. Now, from (9.21), we obtain that

$$\int_0^1 |K_{L,B}(v,x)|\,w(x)\,dx \le 4M\int_0^1 \left(\frac{1+x}{x}\right)^2 e^{-x(v-1)/2}x^2\,dx\,, \tag{9.22}$$

which is bounded for $v\in(1,2]$. To estimate the remaining part of the integral, we write a primitive function $I(v,x)$ of $K_{L,B}(v,x)$ as a Laplace–Stieltjes integral. Thus we have

$$\begin{aligned} I(v,x) &:= -x(v-1)\sum_{n=1}^{\infty} e^{-x(v^n-1)}v^n \\ &= -x\int_0^\infty e^{-xt}\,dF(v,t) \end{aligned}$$

$$= -x^2 \int_0^\infty e^{-xt} F(v,t)\, dt\,, \tag{9.23}$$

where

$$F(v,t) := (v-1) \sum_{n=1}^{I_{t,v}} v^n \tag{9.24}$$

is a step function having jumps of height $(v-1)v^n$ at $t - v^n - 1$, n being a nonnegative integer and $I_{t,v} = \left[\dfrac{\log(t+1)}{\log(v)} \right]$.

Obviously, we have

$$F(v,t) = 0 \qquad \text{for} \quad 0 \leq t < v - 1\,, \tag{9.25}$$

while in addition, we have

$$|F(v,t) \; - \; t| \leq (v-1)(1+t) \qquad \text{for} \quad t \geq v - 1\,. \tag{9.26}$$

Now, to prove (9.26), suppose that $v^n - 1 \leq t < v^{n+1} - 1$ for some nonnegative integer n. In this interval, we have that $F(v,t) = F(v, v^n - 1) = v(v^n - 1)$, and thus,

$$F(v,t) \; - \; t \leq v(v^n - 1) \; - \; (v^n - 1) = (v-1)(v^n - 1) \leq (v-1)t$$

and

$$F(v,t) \; - \; t \geq v(v^n - 1) \; - \; (v^{n+1} - 1) = 1 - v\,.$$

Therefore we get

$$|F(v,t) \; - \; t| \leq \max\{(v-1), (v-1)t\} \leq (v-1)(1+t)$$

and so we have proved (9.26).

From (9.23) and (9.25), we obtain that

$$\begin{aligned} K_{L,B}(v,x) &= -2x \int_0^\infty e^{-xt} F(v,t)\, dt + x^2 \int_0^\infty e^{-xt} t F(v,t)\, dt \\ &= -x^2 \int_{v-1}^\infty e^{-xt} \left(2G(v,t) \; - \; tF(v,t)\right) dt\,, \end{aligned} \tag{9.27}$$

where $G(v,t)$ satisfies $\dfrac{\partial}{\partial t} G(v,t) = F(v,t)$ and $G(v, v-1) = 0$.

From (9.26), we deduce that, for $t \geq v-1$,

$$\left|tF(v,t) - t^2\right| \leq (v-1)(t^2+t),$$

and

$$\left|2G(v,t) - t^2\right| \leq (v-1)(t^2+2t),$$

giving

$$|2G(v,t) - tF(v,t)| \leq 2(v-1)(t^2+2t) \qquad \text{for} \quad t \geq v-1\,. \tag{9.28}$$

Thus, using (9.27), we obtain that

$$\begin{aligned}\int_1^\infty |K_{L,B}(v,x)|\, w(x)\, dx &\leq \int_1^\infty x^2\, dx \int_{v-1}^\infty e^{-xt} 2(v-1)(t^2+2t)\, dt \\ &= 2(v-1)\int_{v-1}^\infty t^2\, dt \int_1^\infty e^{-xt} x^2\, dx \\ &\quad +4(v-1)\int_{v-1}^\infty t\, dt \int_1^\infty e^{-xt} x^2\, dx \\ &< 2(v-1)\int_{v-1}^\infty t^2\, dt \int_1^\infty e^{-xt} x^3\, dx \\ &\quad +8(v-1)\int_{v-1}^\infty \frac{1}{t^2}\, dt \\ &< (12(v-1) + 8(v-1))\int_{v-1}^\infty \frac{1}{t^2}\, dt = 20\,,\end{aligned}$$

which gives (9.12).

9.5.3 Proof of the Lambert theorem

It remains, for the proof of theorem 9.32, first to obtain property (9.13), and then to deduce (9.14). But (9.13) follows at once from (9.27) and (9.28)!

Suppose that $\epsilon > 0$ and that $x_1 < x_2$ are positive numbers which will be chosen suitably (to depend on ϵ). Write

$$L(v) = \left(\int_0^{x_1} + \int_{x_1}^{x_2} + \int_{x_2}^\infty\right) K_{L,B}(v,x) B(x)\, dx$$

$$= L_1(v) + L_2(v) + L_3(v)\,,$$

say. By (9.8), we have that $|B(x)| \le x^2 M'$, where M' is fixed and $x > 0$. Using (9.21), we deduce that

$$|L_1(v)| \le M' \int_0^{x_1} x^2 |K_{L,B}(v,x)|\, dx < \epsilon\,,$$

if $x_1 = x_1(\epsilon)$ is sufficiently small. Further, we deduce from lemma 9.36 and (9.9), that

$$|L_3(v)| \le \sup_{x \ge x_2} |B(x)| \int_1^{\infty} |K_{L,B}(v,x)|\, dx < \epsilon\,,$$

if $x_2 = x_2(\epsilon)$ is sufficiently large. Finally, using (9.13), we deduce that

$$|L_2(v)| \le \sup_{x_1 \le x \le x_2} |B(x)| \int_{x_1}^{x_2} |K_{L,B}(v,x)|\, dx < \epsilon\,,$$

if $|v-1|$ is sufficiently small. Thus the proof is complete.

10

Applications of Borel's methods

Borel's methods in mathematics

10.1 An early application

The bibliography of articles that can be classed as applications of Borel's methods has about 70 entries. Of these, only one predates 1970. That is the article by Cunningham in 1905 **A**[18] entitled
An extension of Borel's exponential method of summation of divergent series applied to linear differential equations.

Here, Cunningham's object was to make more precise the connection between Laplace's integral solutions of linear differential equations with rational coefficients and of rank unity in the neighbourhood of $z = \infty$, and the power series solution with an exponential factor. He noted that Poincaré had developed asymptotic expansions of the integrals, and had found them useful for numerical calculations. However, these asymptotic expansions were of little value in that they did not represent unique analytic functions.

Cunningham proceeded to extend Borel's method to include series of the form:

$$x^{-\rho}\left\{a_0 + \frac{a_1}{x} + \frac{a_2}{x^2} + \dots\right\},$$

where ρ is not an integer.

Starting with the Gamma function, represented as

$$\Gamma(z) = \frac{i}{2\sin(\pi z)}\int_\gamma e^{-\omega}(-\omega)^{z-1}\,d\omega\,,$$

where γ is a contour from infinity, along the real axis, round a small positive circle encircling the origin, and then back to infinity along the real axis.

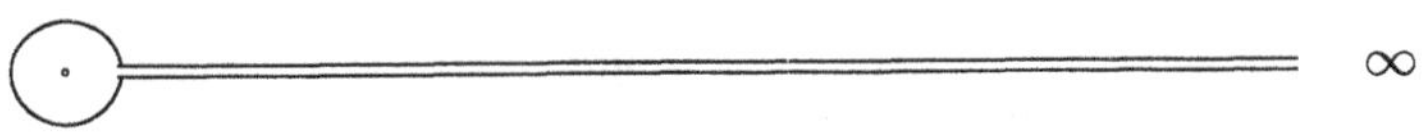

Integrating with respect to ω gives

$$\begin{aligned}\int_\gamma \frac{a_n\omega^{z+n}}{\Gamma(n+z+1)}e^{-\omega}\,d\omega &= (-1)^{-(z+n)}\{-2i\sin(\pi(z+n+1))\}\,a_n \\ &= e^{-\pi i(z+n+1)}\left\{e^{\pi i(z+n+1)} - e^{-\pi i(z+n+1)}\right\}a_n \\ &= a_n(1-e^{-2\pi i z}),\end{aligned}$$

so that

$$a_n = \frac{1}{1-e^{-2\pi i z}}\int_\gamma \frac{a_n\omega^{z+n}}{\Gamma(n+z+1)}e^{-\omega}\,d\omega\,.$$

Consider now the series

$$\mathbf{a} = \sum_{n=0}^{\infty} a_n\,,$$

which may be divergent, and let

$$u_z(\omega) = \sum_{n=0}^{\infty}\frac{a_n\omega^{z+n}}{\Gamma(n+z+1)}\,.$$

Then we have the *formal* equation:

$$\begin{aligned}\frac{1}{1-e^{-2\pi i z}}\int_\gamma e^{-\omega}u_z(\omega)\,d\omega &= \left(\sum_{n=0}^{\infty}\int_\gamma e^{-\omega}\frac{a_n\omega^{z+n}}{\Gamma(n+z+1)}\,d\omega\right)\frac{1}{1-e^{-2\pi i z}} \\ &= \sum_{n=0}^{\infty} a_n.\end{aligned}$$

Further, we obtain another *formal* equation:

$$\begin{aligned}\frac{1}{1-e^{-2\pi i z}}\int_\gamma e^{-\omega}u_z\left(\frac{\omega}{x}\right)\,d\omega &= \sum_{n=0}^{\infty}\frac{a_n}{x^{n+z}} \\ &= x^{-z}\sum_{n=0}^{\infty}a_n x^{-n}\,.\end{aligned}$$

Cunningham then proceeded to demonstrate that this could be used like Borel's method (B').

What is important here is that the Borel sum can be used as an analytic continuation.

10.2 Laplace transforms

In his three volume *Handbuch der Laplace-Transformation*, **T**[12] in volume II, chapter 10, Doetsch discusses the Laplace transforms of entire functions of exponential type.

Given a function $F(t)$ (t real or complex), the Laplace transform is defined by

$$f(s) = \int_0^\infty e^{-st} F(t)\, dt\,.$$

The easiest case occurs when $F(t)$ is entire and given by the power series representation

$$F(t) = \sum_{n=0}^{\infty} \frac{a_n}{n!} t^n\,.$$

It is clear that there must be a connection with Borel's methods of summability. We now obtain, formally, that

$$f(s) = \sum_{n=0}^{\infty} \frac{a_n}{s^{n+1}}\,.$$

If this is convergent, then the region of convergence will be the exterior of a circle. In order to ensure convergence, we require that $F(t)$ satisfies the condition

$$|F(t)| < M\, e^{c|t|}\,,$$

where M and c are positive constants, that is, that F is of exponential type. It is also stated that F is of at most normal type c of order 1. It is also conventional to write

$$f(s) = L\{F\}\,.$$

This leads to

Theorem 10.1. (see Doetsch **T**[12], p. 356, theorem 1). *If $M(r)$ denotes the maximum of $|F(t)|$ for $|t| \leq r$ and ρ is the radius of convergence of the power series $f(s)$, then*

$$\rho = \limsup_{r\to\infty} \frac{\log(M(r))}{r}\,.$$

Thus, $F(t)$ is exactly of normal type ρ of order 1.

From this result, he proceeds to

Theorem 10.2. (see Doetsch **T**[12], p. 358, theorem 2). *If* $F(t) = \sum_{n=0}^{\infty} \frac{a_n\, t^n}{n!}$ *is of exponential type, so that* $f(s) = \sum_{n=0}^{\infty} \frac{a_n}{s^{n+1}}$ *has finite radius of convergence* ρ, *then, in the half plane to the right of the circle of convergence,* $f(s)$ *has a representation as a Laplace transform: that is*

$$f(s) = L\{F\} = \int_0^{\infty} e^{-st} F(t)\, dt \qquad \textit{for } \Re(s) > \rho\,.$$

In section 2, Doetsch gives a generalization of the Laplace transform from the real axis to a ray at angle ϕ to the positive real axis, as

$$L^{(\phi)}\{F\} := \int_o^{\infty(\phi)} e^{-st} F(t)\, dt\,.$$

This leads to a modification of the above theorem, as illustrated by Fig. 10.1 below:

t–plane $\qquad$ s–plane

Fig. 10.1. Ray of integration. $\qquad$ Half plane of convergence.

In, section 5, Doetsch discusses in detail the relations between Borel's integral method of summability applied to a power series and a special case of the Laplace integral transformation. The basis of this work is seen from the Borel transforms of the geometric series, $\sum_{n=0}^{\infty} z^n$. We recall (3.2) that

$$A(x) = \int_0^{x} \left(e^{-x} \sum_{n=0}^{\infty} \frac{a_n x^n}{n!} \right) dx\,.$$

Here we have $a_n = z^n$, so that

$$\sum_{n=0}^{\infty} \frac{a_n x^n}{n!} = \sum_{n=0}^{\infty} \frac{(xz)^n}{n!} = e^{zx} ,$$

yielding that

$$\lim_{x\to\infty} A(x) = \int_0^{\infty} e^{-x} \sum_{n=0}^{\infty} \frac{a_n x^n}{n!}\, dx = \int_0^{\infty} e^{-(1-z)x}\, dx .$$

We note that this integral is convergent for $\Re(z) < 1$ to value $1/(1-z)$. This shows that the geometric series, which is convergent in the unit circle $|z| < 1$, is Borel summable in the half plane $\Re(z) < 1$.

Doetsch has entitled section 5 as *The Borel summability-polygon, the antipolygon, and generalized Borel polygons.* In Chapter 7, we have already made mention of Borel polygons and their generalizations. See theorem 6.4, as well as theorem 131 in Hardy **T**[17].

The main difference inapproach is that Doetsch takes Laplace transforms given by power series of the form

$$f(s) = \sum_{n=0}^{\infty} \frac{a_n}{s^{n+1}} ,$$

so that the region of analyticity is continued into the *exterior* of the convex hull of the singularities, making use of the generalized Laplace integrals $L^{(\phi)}(t)$. See Doetsch's book for details.

10.3 Entire functions and the Borel transform

Here we give a survey of Boas's approach as given in his book **T**[1].

10.3.1 The Phragmén–Lindelöf indicator function

Suppose that $f(z)$ is regular and of exponential type in a closed angle, taken, without loss of generality, to be $|\theta| \le \alpha$. The *Phragmén–Lindelöf indicator function* is defined to be

$$h(\theta) = \limsup_{r\to\infty} \frac{\log\left|f\left(re^{i\theta}\right)\right|}{r} , \qquad |\theta| \le \alpha .$$

A survey of the properties of $h(\theta)$ can be found in chapter 5 of Boas's book **T**[1].

The growth properties of an entire function of exponential type can be described in terms of the geometric properties of a convex set associated with the function. We need the idea of the supporting function.

Let C be a compact, convex set. Project C on to the ray $arg(z) = \theta, 0 \le \theta \le 2\pi$ and call this projection C_θ. Let $k(\theta)$ be the distance from the origin to the most distant point of C_θ in the positive direction. Note that $k(\theta)$ is negative if C_θ lies only on the backwards projection of the ray. Expressed analytically, this is

$$k(\theta) = \max_{x \in C} \Re\left(ze^{-i\theta}\right) = x \cos(\theta) + y \sin(\theta)\,.$$

Since C is a closed set, the line $x \cos(\theta) + y \sin(\phi) - k(\theta) = 0$ contains at least one point of C. Also, C lies entirely to one side of this line when $x \cos(\theta) + y \sin(\theta) \le k(\theta)$.

For $z = r\,e^{i\phi}$, this may be written as $r \cos(\phi - \theta) \le k(\theta)$. This line is called the *supporting line* of C, and $k(\theta)$ is called the *supporting function* of C.

These quantities are not easy to find. Boas gives a good description in section 5.2 of his book **T**[1].

10.3.2 The conjugate indicator diagram

Theorem 10.3. (see Boas, **T**[1], Theorem 5.3.1). *The function*

$$f(z) = \sum_{n=0}^{\infty} a_n\, z^n \tag{10.1}$$

is an entire function of exponential type if and only if

$$F(z) = \sum_{n=0}^{\infty} \frac{a_n\, n!}{z^{n+1}} \tag{10.2}$$

is convergent for some (finite) z.

If the radius of convergence of the series in (10.2) *is* σ, *then* $f(z)$ *is of order* 1 *and either of type* σ *if* $\sigma > 0$, *or of exponential type* 0 *if* $\sigma = 0$.

If $f(z)$ is of order 1 and type τ, then $\limsup_{n\to\infty} |f^{(n)}(0)|^{1/n} = \tau$, and so the radius of convergence of $F(w)$ is τ. On the other hand, if the radius of convergence of $F(w)$ is τ, then the converse follows.

The function $F(z)$ is the **Borel transform** of $f(z)$. It is the Laplace transform of $f(z)$ for z of sufficiently large positive values of $\Re z$, and the analytic continuation of this for other values of z for which it is regular. Indeed, if $z = x + iy$ and $x > \tau$, then

$$\int_0^\infty f(t)e^{-zt}\,dt = \sum_{n=0}^\infty a_n \int_0^\infty t^n e^{-zt}\,dt = \sum_{n=0}^\infty n!\,a_n\,z^{-n-1} = F(z)\,. \quad (10.3)$$

The interchange is justified by Fubini's theorem. Note that

$$\int_0^\infty \sum_{n=0}^\infty |a_n|\,t^n e^{-xt}\,dt = \sum_{n=0}^\infty n!\,|a_n|\,x^{-n-1}\,,$$

where it is clear that the series on the right converges for $x > \tau$, since $\limsup\,(n!|a_n|)^{1/n} = \tau$.

Suppose that $f(z)$ is an entire function of exponential type defined by (10.1), and that D is the intersection of all compact convex sets outside of which $F(z)$ is regular. D is then the smallest closed convex set outside of which $F(z)$ is regular, and is called the *conjugate indicator diagram* of $f(z)$.

Theorem 10.4. (see Boas, **T**[1], Theorem 5.3.5). *Suppose that $f(z)$ is an entire function of exponential type, that D is its conjugate indicator diagram, and that γ is a contour containing D in its interior. Then*

$$f(z) = \frac{1}{2\pi i}\int_\gamma F(w)e^{zw}\,dw\,. \quad (10.4)$$

Boas states that this representation, known as the **Pólya representation**, is one of the most useful tools for investigating entire functions of exponential type. The contour γ can always be taken to be the circle $\Gamma = \{w : |w| = \tau + \epsilon\}$,where τ is the type of $f(z)$ and $\epsilon > 0$. Sometimes, under particular conditions, it can be taken to be the boundary of D.

Note that

$$\begin{aligned}\int_\gamma F(w)e^{zw}\,dw &= \int_\Gamma F(w)e^{zw}\,dw \\ &= \int_\Gamma \left(\sum_{n=0}^\infty n!\,a_n\,w^{-n-1}\sum_{k=0}^\infty \frac{z^k w^k}{k!}\right)dw \\ &= 2\pi i \sum_{n=0}^\infty a_n z^n = 2\pi i\, f(z)\,.\end{aligned}$$

The interchange is justified by the uniform convergence of the series on Γ.

The above enables a more direct characterization of D in terms of $f(z)$.

Theorem 10.5. (see Boas **T**[1], theorem 5.3.7). *The supporting function of D is $h(-\theta)$, where $h(\theta)$ is the indicator function of $f(z)$.*

In other words, $h(\theta)$ is the supporting function of a convex set, the indicator diagram of $f(z)$, and D is the reflection of the indicator diagram in the real axis.

Theorem 10.6. (see Boas **T**[1], theorem 5.3.12). *The extreme points of the conjugate indicator diagram of $f(z)$ are the singular points of the Borel transform $F(z)$ of $f(z)$.*

This is easily shown by considering what happens if an extreme point were not a singular point.

See Boas **T**[1] for a comprehensive survey of properties of the indicator diagram.

Finally, it is possible for the contour γ in the Pólya representation of an entire function of exponential type to be replaced by the boundary ∂D of the conjugate indicator diagram D. The general condition is given in the next theorem. For special cases, see **T**[1].

Theorem 10.7. (see Boas **T**[1], theorem 5.5.1).
Suppose that $\psi(r)$ is a positive valued function satisfying

$$\int_1^\infty \psi(r)\,dr < \infty\,,$$

such that, for all θ,

$$|f(z)| \le \psi(r)e^{rh(\theta)}\,.$$

Then

$$f(z) = \frac{1}{2\pi i}\int_{\partial D} F(w)e^{zw}\,dw\,.$$

10.4 Arithmetical functions

There is a close connection between the Borel mean and density properties of sets of natural numbers. In addition to the Borel mean, some of the analysis relates to the 'circle' methods and some to tauberian results of Hardy and Littlewood in [113] and [114]. These tauberian results are set out in detail in chapter 9 of **T**[17]. In 1980, Borel summability was used by G. Tenenbaum in [278] (see also H. Delange [72]) to investigate the natural density and lower density (sometimes called asymptotic density) of sets of integers. Diaconis and Stein in [73] have demonstrated a connection between Euler density and the Borel method. First the relevant definitions.

Unless otherwise stated, $\mathbb{A}$ is a set of nonnegative integers thought of as a sequence with elements arranged in order of increasing magnitude. Denote by $A(x)$ the 'summing' function defined by:

$$\forall x \in \mathbb{R}, \quad A(x) = \begin{cases} 0 & \text{if } x < 0, \\ \operatorname{card}\{n \in \mathbb{A} : n \le x\} & \text{if } x \ge 0. \end{cases}$$

Definition 10.8. *The set $\mathbb{A}$ is said to have natural density α if $\lim_{x\to\infty} \frac{A(x)}{x}$ exists with value α. The set $\mathbb{A}$ has lower density α if* $\liminf \frac{A(x)}{x} = \alpha$.

Finally, $\mathbb{A}$ has Euler density α with parameter $p \in (0,\, 1)$, or E_p density α if $\lim_{n\to\infty} \sum_{i\in A} b(i,n,p) = \alpha$, where

$$b(i,n,p) = \begin{cases} \binom{n}{k} p^i (1-p)^{n-i} & \textit{for } i \in \{0, \ldots, n\} \\ 0 & \textit{otherwise.} \end{cases}$$

Let E be a nonempty set of prime numbers. For $n \in \mathbb{N}$, denote by $\Omega_E(n)$ and $\omega_E(n)$ the number of prime factors of n which belong to E counted, respectively, with and without regard to multiplicity. When E is the set of all primes write $\Omega(n)$ and $\omega(n)$. Thus, if $n = p_1^{\lambda_1} p_2^{\lambda_2} \cdots p_k^{\lambda_k}$ is the prime decomposition of n, then $\Omega(n) = \lambda_1 + \lambda_2 + \cdots + \lambda_k$ and $\omega(n) = k$.

The functions $\Omega_E(n)$ and $\omega_E(n)$ are 'arithmetical' functions; that is, integer-valued functions defined on $\mathbb{N}$. If g is an arithmetical function let

$$g^{-1}(\mathbb{A}) := \{n \in \mathbb{N} : g(n) \in \mathbb{A}\}.$$

Again, $g^{-1}(\mathbb{A})$ is regarded as an increasing sequence of positive integers.

Tenenbaum, among others, has investigated situations in which $\mathbb{A}$ and $g^{-1}(\mathbb{A})$ have the same natural density. He gives the following result.

Theorem 10.9. *Let E be a set of primes, p, such that $\sum_{p\in E} \frac{1}{p} = \infty$. Set $g = \Omega_E(n)$ or $g = \omega_E(n)$. Then a necessary and sufficient condition on $\mathbb{A}$ such that $g^{-1}(\mathbb{A})$ has a natural density is the existence of an $\alpha \in [0,\, 1]$ such that*

$$\forall c > 0, \quad A(x + c\sqrt{x}) - A(x) = c\alpha\sqrt{x} + \mathbf{o}(\sqrt{x}). \tag{10.5}$$

Moreover, in this case, $\mathbb{A}$ and $g^{-1}(\mathbb{A})$ have density α.

To relate these ideas of density to Borel summability let $\{s_n\}_0^\infty$ be any real (or complex) sequence and set, for $t \geq 0$, $A(t) = \sum_{n \leq t} s_n$. Then

$$s_n \to \alpha(B) \tag{10.6}$$

is the same as

$$\lim_{x\to\infty} e^{-x} \int_0^\infty \frac{x^t}{\Gamma(t+1)} dA(t) = \alpha. \tag{10.7}$$

In the special case when $\{s_n\}_0^\infty$ is the 'characteristic' function of $\mathbb{A}$ in the sense that $s_n = 1$ if $n \in \mathbb{A}$ and $s_n = 0$ otherwise, then the corresponding $A(t)$ is the 'summing' function for $\mathbb{A}$.

To investigate theorem 10.9, Tenenbaum first establishes the following.

Theorem 10.10.
Let E be a set of primes, p, such that $\sum_{p\in E} \frac{1}{p} = \infty$. Set $g = \Omega_E(n)$ or $g = \omega_E(n)$. Then (10.7) *is necessary and sufficient for the sequence $g^{-1}(\mathbb{A})$ to have natural density α.*

Assuming Theorem 10.10 holds, it is clear that 10.9 will be established if conditions (10.5) and (10.7) are equivalent for the functions $A(t)$ generated from the corresponding sequences of 0s and 1s.

Tenenbaum shows a bit more. Let $\mathbb{V}$ denote the family of complex valued functions of bounded variation on all bounded intervals.

Theorem 10.11.
Let $A \in \mathbb{V}$. Then (10.5) *implies* (10.7).

Theorem 10.12.
Let $A \in \mathbb{V}$. If there exists a complex number β and a function $B(x)$ satisfying

$$\forall c > 0, \quad B(x + c\sqrt{x}) - B(x) = c\beta\sqrt{x} + \mathbf{o}(\sqrt{x}),$$

such that the function $A(x)+B(x)$ has monotonic real and imaginary parts, then (10.7) *implies* (10.5).

Corollary 10.13.
Let $\{s_n\}$ be a sequence of real numbers. If there exists a real number λ such that $s_n \geq -\lambda$ for large enough n, and if $s_n \to \alpha(B)$ then

$$\forall c > 0, \quad \sum_{x\leq n<c\sqrt{x}} s_n = \alpha x + \mathbf{o}(\sqrt{x}). \tag{10.8}$$

Tenenbaum notes that when $A(t)$ is the summing function corresponding to $\{s_n\}$ then the corollary, with $B(x) = \lambda[x]$, together with theorems 10.10 and 10.11, imply theorem 10.9.

We content ourselves here with giving Tenenbaum's proof of theorem 10.10. Let $K(x,t)$ denote the kernel in equation (10.7). That is, $K(x,t) = e^{-x}\dfrac{x^t}{\Gamma(t+1)}$.

Lemma 10.14.

1. *For $x > 0$ and $|t-x| \le x^{\frac{2}{3}}$,*

$$K(x,t) = \frac{1}{\sqrt{2\pi x}} e^{-\frac{(t-x)^2}{2x}} \left(1 + \mathbf{O}\left(\frac{1+|t-x|}{x}\right) + \mathbf{O}\left(\frac{|t-x|^3}{x^2}\right)\right).$$

2. *For $x > 0$ and $|t-x| \le \dfrac{x}{2}$,*

$$\frac{\partial K}{\partial t}(x,t) = \mathbf{O}\left(\frac{1+|t-x|}{x} K(x,t)\right).$$

3. *For all $x > 0$, $K(x,t)$ as a function of t is positive and attains its absolute maximum at a point $t(x)$ in the interval $[x-1,\, x]$; it is increasing on $[0,\, t(x))$ and decreasing on $(t(x),\, \infty)$.*

4. *For $\dfrac{1}{2} < \nu < \dfrac{2}{3}$ and $\epsilon > 0$,*

$$\int_{|t-x|>x^{\nu}} K(x,t)dt = \mathbf{O}\left(e^{-x^{2\nu-1-\epsilon}}\right).$$

5. *As x tends to infinity,*

$$\int_0^{\infty} K(x,t)dt = 1 + \mathbf{O}(e^{-x}).$$

As in theorem 10.9, let E be a set of primes, p, such that $\sum_{p\in E} \dfrac{1}{p} = \infty$. and set $g = \Omega_E(n)$ or $g = \omega_E(n)$. The following notations are used in the next lemma and in the proof of theorem 10.10. For $x > 0$ and m a positive integer,

$$y(x) = \sum_{\substack{p \le x \\ p \in E}} \frac{1}{p},$$

and

$$N(m,x) = \mathrm{card}\{n \le x \colon g(n) = m\}.$$

Lemma 10.15. (see Halász, [98], theorem 1). *Suppose* $0 < \delta \le 1$. *If* $y(x) \to \infty$ *then*

$$N(m,x) = xK(y,m)\left(1 + \mathbf{O}\left(\frac{|m-y|}{y}\right) + \mathbf{O}\left(\frac{1}{\sqrt{y}}\right)\right), \tag{10.9}$$

uniformly for $\delta \le \dfrac{m}{y} \le 2 - \delta$ *and* $y \ge 2$.

Proof of theorem 10.10 Write $\mathbf{B} = g^{-1}(\mathbf{A})$ and let $B(x)$ represent its summing function. Then $B(x) = \sum_{m \in \mathbf{A}} N(m,x)$.

Fix $\nu \in \left(\frac{1}{2}, \frac{2}{3}\right)$ and write

$$B(x) = \sum_{\substack{m \in \mathbf{A} \\ |m-y| \le y^\nu}} N(m,x) + \mathbf{O}\left(\sum_{|m-y| \ge y^\nu} N(m,x)\right). \tag{10.10}$$

When $\mathbf{A}$ is the sequence of all positive integers one has $B(x) = [x]$. But using (10.9) and lemma (10.14) give

$$\begin{aligned}
&\sum_{|m-y| \le y^\nu} N(m,x) \\
&= x\left(1 + \mathbf{O}\left(\frac{1}{y^{1-\nu}}\right)\right) \sum_{|m-y| \le y^\nu} K(y,m) \\
&= x\left(1 + \mathbf{O}\left(\frac{1}{y^{1-\nu}}\right)\right)\left(\int_{|t-y| \le y^\nu} K(y,t)dt + \mathbf{O}(\frac{1}{\sqrt{y}})\right) \\
&= x\left(1 + \mathbf{O}\left(\frac{1}{y^{1-\nu}}\right)\right).
\end{aligned}$$

Hence for any sequence $\mathbb{A}$ the remaining term of (10.10) is dominated by

$$[x] - x\left(1 + \mathbf{O}\left(\frac{1}{y^{1-\nu}}\right)\right) = \mathbf{o}(x).$$

Therefore, from (10.10) and using (10.9) we have

$$\begin{aligned}\frac{B(x)}{x} &= \left(1 + \mathbf{O}\left(\frac{1}{y^{1-\nu}}\right)\right) \sum_{|m-y|\leq y^{\nu}} K(y,m) + \mathbf{o}(1) \\ &= (1 + \mathbf{o}(1))\left(\int_0^{\infty} K(y,t)dA(t) + \mathbf{o}(1)\right) + \mathbf{o}(1).\end{aligned}$$

Hence, $\lim_{x\to\infty} \dfrac{B(x)}{x} = \alpha$ if and only if (10.7) holds. This completes the proof.

In the paper by Diaconis and Stein, which predates that of Tenenbaum, the main result is the following. Here, $\mathbf{A}$, $A(t)$, and $\{s_n\}_0^{\infty}$ (the 'characteristic' function of $\mathbf{A}$) all retain their earlier meaning.

Theorem 10.16.
The following are equivalent:

$$\mathbb{A} \textit{ has } E_p \textit{ density } \alpha\,, \tag{10.11}$$

$$s_n \to \alpha\,(B)\,, \tag{10.12}$$

$$\forall \epsilon > 0, \quad \lim_{n\to\infty} \frac{1}{\epsilon\sqrt{n}} \sum_{\substack{i\in\mathbb{A} \\ n\leq i<n+\epsilon\sqrt{n}}} 1 \;=\; \alpha\,. \tag{10.13}$$

First, note that condition (10.13) is the earlier condition (10.5). Moreover, Euler density E_p is identical to Euler summability E_q where $p = \dfrac{1}{1+q}$. Hence, that (10.11) implies (10.12) follows immediately from theorem 2.28. Diaconis and Stein show that (10.12) implies (10.13) by applying Wiener's tauberian theorem (see **T**[17], theorem 220). This implication also follows from corollary 10.13 above. That condition (10.13) implies Euler density α and, hence, that the notion is independent of p is proved directly by Diaconis and Stein. Their proof uses the central limit theorem (see Feller, **T**[15], for example). We omit the details.

Recent applications in mathematical physics

As mentioned above, the bibliography of articles that can be classed as applications of Borel's methods has about 70 entries. There is also one textbook and two volumes in the Springer–Verlag Lecture Notes in Physics series.

10.5 Basic theory

Reed and Simon **T**[25], in the fourth volume of their series entitled *Methods of Modern Mathematical Physics*, indicate the basis of the recent applications of Borel's methods to some problems in mathematical physics.

Definition 10.17. *Suppose that $F(z)$ is analytic in a sector $\mathbb{S} = \{z : 0 < |z| < \alpha; |\arg(z)| < \frac{\pi}{2} + \epsilon\}$ and satisfies the following condition: there exist constants C and γ such that*

$$\left| f(z) - \sum_{n=0}^{N} a_n z^n \right| \leq C\gamma^{N+1}(N+1)!|z|^{N+1}$$

for all N and all $z \in \mathbb{S}$.

Then we say that $f(z)$ obeys a **strong asymptotic condition** *and has $\sum_{n=0}^{\infty} a_n z^n$ as its* **strong asymptotic series**.

Reed and Simon then proved that such strong asymptotic series uniquely determine an analytic function.

Theorem 10.18. (see Reed and Simon **T**[25], theorem XII.19). *If the series $\sum_{n=0}^{\infty} a_n z^n$ is a strong asymptotic series for two analytic functions f and g, then $f = g$.*

They then proceed to give Watson's theorem [305] in the following form:

Theorem 10.19. (see Reed and Simon **T**[25], theorem XII.21). *Suppose that $f(x)$ has $\sum_{n=0}^{\infty} a_n z^n$ as its strong asymptotic series in a sector $\mathbb{S} = \{z : 0 < |z| < \alpha; |\arg(z)| < \frac{\pi}{2} + \epsilon\}$. Then $a(z) = \sum_{n=0}^{\infty} \frac{a_n}{n!} z^n$ is analytic in some disk centred on the origin and*

(a) *$a(z)$ has an analytic continuation into the sector* $\{z : |\arg(z)| < \epsilon\}$*;*

(b) *if* $|z| < \alpha$ *and* $|\arg(z)| < \epsilon$, *then* $\int_0^\infty e^{-x}|a(zx)|\, dx < \infty$*;*

(c) *if* $|z| < \alpha$ *and* $|\arg(z)| < \epsilon$, *then* $f(z) = \int_0^\infty e^{-x}|a(zx)|\, dx$.

Reed and Simon note that they are making use of Borel's (integral) summability method. Again, what is important here is that the Borel sum can be used as an analytic continuation. This is the basis for the use of summability methods of the Borel type in some research papers in mathematical physics since 1970.

Note that Reed and Simon have a minor error in their definition of a regular summability method **T**[25], page 63. They state that a summability method *is called regular if, whenever* $\sum a_n$ *is* **absolutely** *convergent to* σ*, the procedure is applicable and yields* σ *as sum.* The standard definition of regularity requires only that the original series is convergent to σ.

It seems strange that the essence of Cunningham's paper **A**[18], that Borel's means can be used to give an analytic continuation that can be found explicitly, was not rediscovered until 1970, a gap of 65 years.

The reader is referred to the 'applied bibliography' for a list of recent applications in theoretical physics.

References

This chapter contains many more items than those to which there is a reference in the preceding chapters. The extra references are included to give as full a listing as possible in the knowledge that they will be useful to those interested in Borel's methods of summability. Although this listing is comprehensive, we do not claim that it is complete. It is, though, the result of many hours of bibliographic searching.

Bibliography

TEXTBOOKS, THESES, LECTURE NOTES

1. R.P. Boas. *Entire Functions.* Academic Press, New York, 1954.

2. É. Borel. *Leçons sur les séries divergentes.* Gauthier-Villars, Paris, 1901.

3. É. Borel. *Leçons sur les séries divergentes.* Gauthier-Villars, Paris, 2nd edition, 1928. Collection de monographies sur la théorie des fonctions.

4. É. Borel. *Lectures on Divergent Series.* Los Alamos Scientific Laboratory, University of California, Los Alamos, N.M., 1975. This is a translation, by L. Critchfield and A. Vakar, of Leçons sur les séries divergentes.

5. L.S. Bosanquet. *The History and Development of the Theory of Divergent Series and Integrals.* Unpublished Lecture Notes, University of Utah, 1964–65.

6. L.S. Bosanquet. *An Introduction to Functional Analysis, with Applications to Summability Theory.* Unpublished Lecture Notes, University of Western Ontario, 1970.

7. T.J.I'A. Bromwich. *An Introduction to the Theory of Infinite Series.* Cambridge University Press, London, 1907.

8. P. Collet and J-P. Eckmann. *A Renormalization Group Analysis of the Hierarchical Model in Statistical Mechanics.* Volume 74 of *Lecture Notes in Physics*, Springer–Verlag, Berlin-Heidelberg-New York, 1978.

9. R.G. Cooke. *Infinite Matrices and Sequence Spaces.* Macmillan, London, 1950.

10. P. Dienes. *The Taylor Series.* Oxford University Press, Oxford, 1931.

11. G. Doetsch. *Eine neue Verallgemeinerung der Borelschen Summabilitätstheorie der divergenten Reihen.* Diss., Göttingen, 1920.

12. G. Doetsch. *Handbuch der Laplace–Transformation.* Birkhäuser, Basel, 1956.

13. N. Dunford and J.T. Schwartz. *Linear Operators*, Part I. Interscience Publishers, New York, 1958.

14. J-P. Eckmann and P. Wittwer. *Computer Methods and Borel Summability applied to Feigenbaum's Equation.* Volume 85 of *Lecture Notes in Physics*, Springer–Verlag, Berlin-Heidelberg-New York, 1985.

15. W. Feller *An Introduction to Probability Theory and its Applications.* Wiley, New York, 3rd edition, 1968.

16. W.B. Ford. *Studies on Divergent Series and Summability.* Chelsea Publishing Company, New York. Reprint of 1960 edition.

17. G.H. Hardy. *Divergent Series.* Oxford University Press, Oxford, 2nd edition, 1949.

18. G.H. Hardy. and M. Riesz *The General Theory of Dirichlet's Series.* Cambridge Mathematical Tract No. 18, Cambridge, 1915.

19. K. Knopp (R.C.H. Young, Translator). *Theory and Application of Infinite Series.* Hafner, New York, 1971.

20. J.E. Littlewood *Lectures on the Theory of Function.* Oxford University Press, Oxford, 1943.

21. A. Peyerimhoff. *Lectures on Summability.* Volume 107 of *Lecture Notes in Mathematics*, Springer–Verlag, Berlin-Heidelberg-New York, 1969.

22. H.R. Pitt. *Tauberian Theorems.* Volume 2 of *Tata Institute for Fundamental Research Monographs on Mathematics and Physics*, Oxford University Press, Oxford, 1958.

23. R.E. Powell and S.M. Shah *Summability Theory and Applications.* Van Nostrand Reinhold, London, 1972.

24. R.E. Powell and S.M. Shah *Summability Theory and Applications.* Prentice–Hall of India, New Delhi, 1988. Revised edition.

25. M. Reed and B. Simon. *Methods of Modern Mathematical Physics, Volume IV. Analysis of Operators.* Academic Press, New York, 1978.

26. G.F. Simmons. *Calculus Gems.* McGraw Hill, New York, 1992.

27. K.R. Stromberg. *An Introduction to Classical Real Analysis.* Wadsworth, Belmont, California, 1981.

28. E.C. Titchmarsh. *The Theory of Functions.* Oxford University Press, Oxford, 2nd edition, 1939.

29. R. Wais. Das Taylorsche Summierungsverfahren. Dissertation, Tübingen, 1935, 58 pp.

30. E.T. Whittaker and G.N. Watson. *A Course of Modern Analysis.* Cambridge University Press, Cambridge 4th edition, 1927.

31. D.V. Widder *An Introduction to Transform Theory.* Academic Press, New York, 1971.

32. K. Zeller and W. Beekmann *Theorie der Limitierungsverfahren.* Springer–Verlag, Berlin-Heidelberg-New York, 2nd edition, 1970.

Bibliography

ARTICLES

1. R.P. Agnew. The effects of general transformations on oscillations of sequences of functions. *Trans. Amer. Math. Soc.*, **33** (1931) 411–424.

2. R.P. Agnew. Borel transforms of Tauberian series. *Math. Z.*, **67** (1957) 51–62.

3. R.P. Agnew. Relations among the Lototsky, Borel and other methods for evaluation of series. *Michigan Math. J.*, **6** (1959) 363–371.

4. M. Akita and T. Kano. On the Borel summability of $\sum_{n=1}^{\infty} n^{-\alpha} \exp(in^{\beta}\theta)$. *Proc. Japan Acad. Ser. A Math. Sci.*, **59** (1983) 270–273.

5. B.M. Bajšanski. Sur une classe générale de procédés de sommations du type d'Euler-Borel. *Acad. Serbe Sci. Publ. Inst. Math.*, **10** (1956) 131–152.

6. B.M. Bajšanski. Une classe générale de procédés de sommations du type d'Euler-Borel et leur application au prolongement analytique. *Srpska Akad. Nauka, Zbornik Radova*, **63** (1959) 1–35.

7. E.W. Barnes. A memoir of integral functions. *Proc. Royal Soc. London* , **69** (1902) 121–125.

8. E.W. Barnes. The generalization of the Maclaurin sum formula, and the range of its applicability. *Quart. J. Math. Oxford*, **35** (1903) 175–188.

9. V.A. Berekašvili. Borel summation of double series. *Soobščeniya Akad. Nauk Gruzin. SSR*, **14** (1953) 193–196. Russian.

10. F. Bernstein. Die, Übereinstimmung derjenigen beiden Summationsverfahren einer divergenten Reihe, weiche von T.E. Stieltjes und É. Borel herrühren. *Deutsche Math.-Ver.*, **28** (1919) 50–63.

11. V. Bernstein. Sur une généralisation de la méthode de sommation exponentielle de M. Borel. *C. R. Acad. Sci. Paris Sér. I Math.*, **194** (1932) 1887–1889.

12. V. Bernstein. Sulle direzioni di Borel di funzioni olomorfe. *Ann. Mat. pura appl., IV.s.*, **12** (1934) 173–196.

13. W. Biegert. Über Tauber-Konstanten beim Borel-Verfahren. *Math.-Z.*, **92** (1966) 331–339.

14. W. Biegert. Tauber-Konstanten zu verschiedenen Tauber-Bedingungen beim Borel-Verfahren. *Indian J. Math.*, **9** (1967) 25–36. English summary.

15. N.H. Bingham. On Valiron and circle convergence. *Math. Z.*, **186** (1984) 273–286.

16. N.H. Bingham. On Borel and Euler summability. *J. London Math. Soc.* (2), **29** (1984) 141–146.

17. N.H. Bingham. Tauberian theorems for summability methods of random walk type. *J. London Math. Soc.* (2), **30** (1984) 281–287.

18. A. Birkholc. On generalized power methods of limitation. *Bull. Acad. Polon. Sci. Sér. Sci. Math. Astronom. Phys.*, **13** (1965) 323–327.

19. A. Birkholc. On generalized power methods of limitation. *Studia Math*, **27** (1966) 213–245.

20. J. Boos. Convexity theorems for the circle methods of summability. *J. Comput. Appl. Math.*, **40** (1992) 151–155.

21. É. Borel. Sur la sommation des séries divergentes. *C. R. Acad. Sci. Paris Sér. I Math.*, **121** (1895) 1125–1127.

22. É. Borel. Applications de la théorie des séries divergentes sommables. *C. R. Acad. Sci. Paris Sér. I Math.*, **122** (1896) 805–807.

23. É. Borel. Fondements de la théorie des séries divergentes sommables. *Journ. de Math*, **2** (1896) 103–122.

24. É. Borel. Sur les séries de Taylor. *C. R. Acad. Sci. Paris Sér. I Math.*, **123** (1896) 1051–1052.

25. É. Borel. Sur la généralisation de la notion de limite et sur l'extension aux séries divergentes sommables du théorème d'Abel sur les séries entières. *C. R. Acad. Sci. Paris Sér. I Math.*, **122** (1896) 73–74.

26. É. Borel. Sur la région de sommabilité d'un developpement de Taylor. *C. R. Acad. Sci. Paris Sér. I Math.*, **123** (1896) 548–549.

27. É. Borel. Sur les séries de Taylor admettant leur cercle de convergance comme coupure. *Journ. de Math*, **2** (1896) 441–451.

28. É. Borel. Mémoire sur les séries divergentes. *Ann. de l'Éc. Norm.*, **16** (1899) 9–131, 132–136.

29. É. Borel. Sur le calcul des séries de Taylor à rayon de covergence nul. *C. R. Acad. Sci. Paris Sér. I Math.*, **128** (1899) 1281–1283.

30. É. Borel. Les séries absolument sommables, les séries (M), et le prolongement analytique. *C. R. Acad. Sci. Paris Sér. I Math.*, **131** (1900) 830–832.

31. É. Borel. Sur le prolongement analytique de la série de Taylor. *S.M.F. Bull.*, **28** (1900) 200.

32. É. Borel. Le prolongement analytique et les séries sommables. *Math. Ann.*, **55** (1901) 74–80.

33. É. Borel. Sur la méthode de sommation exponentielle des séries divergentes. *Matem. Tidsskr.*, **B** (1924) 40–41.

34. D. Borwein. On a scale of Abel-type summability methods. *Proc. Cambridge Phil. Soc.*, **53** (1957) 318–322.

35. D. Borwein. On methods of summability based on power series. *Proc. Royal Soc. Edinburgh*, **64** (1957) 342–349.

36. D. Borwein. A logarithmic method of summability. *J. London Math. Soc.* (2), **33** (1958) 212–220.

37. D. Borwein. On Borel-type methods of summability. *Mathematika*, **5** (1958) 128–133.

38. D. Borwein. Theorems on some methods of summability. *Quart. J. Math. Oxford* (2), **9** (1958) 310–316.

39. D. Borwein. On methods of summability based on integral functions. I. *Proc. Cambridge Phil. Soc.*, **55** (1959) 23–30.

40. D. Borwein. On methods of summability based on integral functions. II. *Proc. Cambridge Phil. Soc.*, **56** (1960) 125–131.

41. D. Borwein. On moment constant methods of summability. *J. London Math. Soc.*, **35** (1960) .

42. D. Borwein. Relations between Borel-type methods of summability. *J. London Math. Soc.* (1), **35** (1960) 65–70.

43. D. Borwein. A Tauberian theorem for Borel-type methods of summability. *Canad. J. Math.*, **21** (1969) 740–747.

44. D. Borwein. On absolute Borel-type methods of summability. *Proc. Amer. Math. Soc.*, **24** (1970) 85–89.

45. D. Borwein. A Tauberian theorem concerning Borel-type and Riesz summability methods. *Canad. Math. Bull.*, **35** (1992) 14–20.

46. D. Borwein and T. Markovich. A Tauberian theorem concerning Borel-type and Cesàro methods of summability. *Canad. J. Math.*, **40** (1988) 228–247.

47. D. Borwein and T. Markovich. Cesáro and Borel-type summability. *Proc. Amer. Math. Soc.*, **103** (1988) 1108–1112.

48. D. Borwein and T. Markovich. Weighted means and summability by the circle and other methods. *J. Approx. Theory*, **68** (1992) 49–55.

49. D. Borwein and I.J.W. Robinson. A Tauberian theorem for Borel-type methods of summability. *J. Reine Angew. Math.*, **273** (1975) 153–164.

50. D. Borwein and B.L.R. Shawyer. On Borel-type methods. *Tôhoku Math. J.* (2), **18** (1966) 283–298. See correction.

51. D. Borwein and B.L.R. Shawyer. On Borel-type methods. II. *Tôhoku Math. J.* (2), **19** (1967) 232–237.

52. D. Borwein and E. Smet. Tauberian theorems for Borel-type methods of summability. *Canad. Math. Bull.*, **17** (1974) 167–173. See addendum.

53. D. Borwein and E. Smet. Addendum to 'Tauberian theorems for Borel-type methods of summability'. *Canad. Math. Bull.*, **19** (1976) 245–246.

54. D. Borwein and E. Smet. Tauberian theorems for strong and absolute Borel-type methods of summability. *Canad. Math. Bull.*, **20** (1977) 161–172.

55. D. Borwein and B. Watson. Tauberian theorems on a scale of Abel-type summability methods. *J. Reine Angew. Math.*, **298** (1978) 1–7.

56. D. Borwein and B. Watson. On the relation between the logarithmic and Borel-type summability methods. *Canad. Math. Bull.*, **24** (1981) 153–159.

57. D. Borwein and B. Watson. Tauberian theorems between the logarithmic and Abel-type summability methods. *Pacific J. Math.*, **101** (1982) 11–23.

58. G. Bouligand. Sur la comparaison de certains procédés de sommation des séries divergentes. *Enseignement Math.*, **26** (1927) 15–27.

59. U. Broggi. Su di un'applicazione del metodo di sommazione di Borel. *Atti Accad. naz. Lincei, Rend., VI.s.*, **19** (1934) 378–382.

60. A. Buhl. Application du procédé de sommation de M.E. Borel aux séries trigonometriques généralisées. *C. R. Acad. Sci. Paris Sér. I Math.*, **143** (1906) 445–446.

61. A. Buhl. Sur une extension de la méthode de sommation de M. Borel. *C. R. Acad. Sci. Paris Sér. I Math.*, **144** (1907) 710–712.

62. A. Buhl. Sur la représentation des fonctions méromorphes. *Acta Math.*, **35** (1911) 73–95.

63. M.V. Čelidze. Summability of double series by the Borel method. *Sakharth. SSR Mecn. Akad. Moambe*, **66** (1972) 271–274. Russian (Georgian and English summaries).

64. E. Cesàro. Sur la multiplication des séries. *Bull. des Sci. Math.* (2), **14** (1890) 114–120.

65. Y.S. Chow Delayed sums and Borel summability of independent, identically distributed random variables. *Bull. Inst. Math. Acad. Sinica*, **1** (1973) 207–220.

66. R.G. Cooke. On mutual consistency and regular T-limits. *Proc. London Math. Soc.* (2), **41** (1936) 113–125.

67. R.G. Cooke. An extension of some recent results on mutual consistency and regular T-limits. *J. London Math. Soc.*, **12** (1937) 98–105.

68. A. Costabel. Sur le prolongement analytique d'une fonction méromorphe. *L'Ens. Math.*, **10** (1908) 377–390.

69. V.F. Cowling. On Borel summability. *J. London Math. Soc.* (1), **31** (1956) 369–373.

70. N.A. Davydov. [M.O. Davidov]. A property of the Borel method of summation of series and Tauberian theorems. Over-convergence of power series and singular points of analytic functions. *Kalinin. Gos. Ped. Inst. Uč. Zap.*, **26** (1958) 57–82. Russian.

71. N.A. Davydov [M.O. Davidov] and V. Lototskiĭ [V.A. Lotockiĭ]. On the kernel of Borel means. *Ukrain. Mat. Ž.*, **32** (1980) 11–18, 141. Russian.

72. H. Delange Sur la distribution des entiers ayant certains propriétés. *Ann. Sci. Ec. Norm. Sup.* (3e *Série*), **73** (1956) 15–74.

73. P. Diaconis and C. Stein. Some Tauberian theorems related to coin tossing. *Annals of Probability*, **6** (1978) 483–490.

74. G. Doetsch. Über die Summabilität von Potenzreihen auf dem Rande des Borelschen Summabilitätspolygons. *Math. Ann.*, **84** (1921) 245–251.

75. G. Doetsch. Über den Zusammenhang zwischen Abelscher und Borelscher Summabilität. *Math. Ann.*, **104** (1931) 403–414.

76. A. Durañona. Sobre producto de series sumables Borel. *Contribución La Plata*, **4** (1928) 435–449.

77. P. Erdös. On a Tauberian theorem for Euler summability. *Acad. Serbe Sci. Publ. Inst. Math.*, **4** (1952) 51–56.

78. P. Erdös. On a high-indices theorem in Borel summability. *Acta Math. Acad. Sci. Hungar.*, **7** (1956) 265–281. Russian summary.

79. A. Evans. The application of complex variable methods to Tauberian theorems. *J. London Math. Soc.*, **28** (1953) 94–102. Russian summary.

80. E. Fabry. Sur les points singuliers d'une fonction donn/'ee par son déveoppement en série et sur l'impossibilité du prolongement analytique dans les cas très généraux. *Annales scientifiques de l'École normale supérieure* (3), **13** (1896) 107–114.

81. G. Faulhaber. Aquivalenzsatze für die Kreisverfahren der Limitierungstheorie. *Math. Z.*, **66** (1956) 34–52.

82. T. Fort. Borel summability and Lambert series. *Amer. J. Math.*, **61** (1939) 397–402.

83. M. Fujiwara. Ein Satz über die Borelsche Summation. *Tôhoku Math. J.* (1), **17** (1920) 339–343.

84. D. Gaier. Zur Frage der Indexverschiebung beim Borel-Verfahren (On the problem of the translativity of the Borel exponential method). *Math. Z.*, **58** (1953) 453–455.

85. D. Gaier. On modified Borel methods. *Proc. Amer. Math. Soc.*, **6** (1955) 873–879.

86. D. Gaier. Über die Äquivalenz der $|B_k|$-Verfahren. *Math. Z.*, **64** (1956) 183–191.

87. D. Gaier. Der allgemeine Lückenumkehrsatz für das Borel-Verfahren. *Math. Z.*, **88** (1965) 410–417.

88. D. Gaier. On the coefficients and the growth of gap power series. *SIAM J. Numer. Anal.*, **3** (1966) 248–265.

89. M.K. Ganapathy. A note on (B, α, β) summability. *Bull. Calcutta Math. Soc.*, **74** (1982) 299–307.

90. M.K. Ganapathy. Borel-type transforms of Tauberian series. *Bull. Calcutta Math. Soc.*, **82** (1990) 22–26.

91. A. Ganguly and R.K. Jain. On some relations between strong and ordinary almost Borel-type methods of summability. *J. Indian Acad. Math.*, **3** (1981) 37–39.

92. V. Garten. Über den Einflußendlich vieler Änderungen auf das Borelsche Limitierungsverfahren. *Math. Z.*, **40** (1936) 756–759.

93. V. Garten and K. Knopp. Ungleichungen zwischen Mittelwerten von Zahlenfolgen und Funktionen. *Math. Z.*, **42** (1937) 365–388.

94. W. Gawronski, H. Siebert, and R. Trautner. An inclusion relation for Abel, Borel, and Lambert summability. *Canad. J. Math.*, **32** (1980) 695–702.

95. W. Gawronski and R. Trautner. Verschärfung eines Satzes von Borel-Okada über Summierbarkeit von Potenzreihen. *Period. Math. Hungar.*, **7** (1976) 201–211.

96. I.J. Good. Relations between methods of summation of series. *Math. Proc. Cambridge Philos. Soc.*, **38** (1942) 144–165.

97. I.J. Good. On the regularity of moment methods of summation. *J. London Math. Soc.*, **19** (1944) 141–143.

98. G. Halász. On the distribution of additive and the mean values of multiplicative arithmetic functions. *Studia Sci. Math. Hung.*, **6** (1971) 211–233.

99. G. Halász. Remarks to a paper of D. Gaier on gap theorems. *Acta Sci. Math. (Szeged)*, **28** (1967) 311–322.

100. L. Hanni. Über Borels Verallgemeinerung des Grenzbegriffes. *Monatsh. Math.*, **12** (1901) 265–289.

101. L. Hanni. Zurückführung der allgemeinen Mittelbildung Borels auf Mittag-Lefflers n-fach unendliche Reihen. *Monatsh. Math.*, **14** (1903) 105–124.

102. L. Hanni. Über den Beziehungen zwischen den Darstellung eines eindeutigen Zweiges einer monogenen Funktion durch Herrn Mittag-Leffler, der Methode der Mittelwerte des Herrn Borel, und der Transformation des Herrn Lindelöf. *Acta Math.*, **29** (1904) 25–58.

103. G.H. Hardy. Researches in the theory of divergent series and divergent integrals. *Quart. J. Math. Oxford*, **35** (1903), 22–26.

104. G.H. Hardy. On differentiation and integration of divergent series. *Trans. Cambridge Philos. Soc.*, **19** (1904), 297–321.

105. G.H. Hardy. Further researches in the theory of divergent series and divergent integrals. *Cambridge Philos. Soc. Trans.*, **21** (1908), 1–48.

106. G.H. Hardy. The application to Dirichlet's series of Borel's exponential method of summation. *Proc. London Math. Soc.* (2), **8** (1910) 277–294.

107. G.H. Hardy. Notes on some points in the integral calculus. XXXI. The uniform convergence of Borel's integral. *Messenger*, **40** (1911) 161–165.

108. G.H. Hardy. The uniform convergence of Borel's integral. *Messenger*, **40** (1911), 161–165.

109. G.H. Hardy. Notes on some points in the integral calculus. XXXVII. On the region of convergence of Borel's integral. *Messenger*, **43** (1913) 22–24.

110. G.H. Hardy. On the summability of series by Borel's and Mittag-Leffler's methods. *J. London Math. Soc.* (1), **9** (1934) 153–157.

111. G.H. Hardy and J.E. Littlewood. The relations between Borel's and Cesàro's methods of summation. *Proc. London Math. Soc.* (2), **11** (1912) 1–16.

112. G.H. Hardy and J.E. Littlewood. Contributions to the arithmetic theory of series. *Proc. London Math. Soc.* (2), **11** (1913) 411–478.

113. G.H. Hardy and J.E. Littlewood. Theorems concerning the summability of series by Borel's exponential method. *Rend. Circ. Mat. Palermo* (2), **41** (1916) 36–53.

114. G.H. Hardy and J.E. Littlewood. On a Tauberian theorem for Lambert's series and some fundamental theorems in the analytic theory of numbers. *Proc. London Math. Soc.* (2), **19** (1919) 21–29.

115. G.H. Hardy and J.E. Littlewood. Notes on the theory of series. XXII. On the Tauberian theorem for Borel summability. *J. London Math. Soc.* (1), **18** (1943) 194–200.

116. G. Hayashi. Einige Sätze über Borelsche und Abelsche Summierungen der divergenten Reihen. *Tôhoku Math. J.* (1), **37** (1933) 164–168.

117. G. Hayashi. Bemerkungen über die Borelschen- und Abelschen-Summierungsverfahren der Divergenten Reihen. *Sci. Rep. Tohoku Univ., I.s.*, **23** (1935) 920–926.

118. O. Henriksson. Über die Hausdorffschen Limitierungsverfahren, die schwächer sind, als das Abelsche. *Math. Z.*, **39** (1935) 501–510.

119. E. Hensz. On the perfectness of the Cesàro, Borel and Abel limitation methods of sequences. *Bull. Acad. Polon. Sci. Sér. Sci. Math. Astronom. Phys.*, **23** (1975) 1079–1081. Russian summary.

120. L. Hoischen. Beiträge zur Limitierungstheorie. *Mitt. math. Sem. Geißen*, **59** (1962) 56 pp.

121. L. Hoischen. An inclusion theorem for Abel and Lambert summability. *J. London Math. Soc.* (1), **42** (1967) 591–594.

122. L. Hoischen. Some inclusion theorems for generalized Abel and Borel summability. *J. London Math. Soc.* (1), **42** (1967) 229–234.

123. L. Hoischen. Über die Wirkfelder verallgemeinerter Abelscher Limitierungsververfahren. *Acta Math. Acad. Sci. Hung.*, **20** (1969) 149–157.

124. W.A. Hurwitz. Report on topics in the theory of divergent series. *Bull. Amer. Math. Soc.*, **28** (1922) 17–36.

125. J.M. Hyslop. The generalization of a theorem on Borel summability. *Proc. London Math. Soc.* (2), **41** (1936) 243–256.

126. K. Ikeno. Summability methods of Borel type and Tauberian series. *Tôhoku Math. J.* (2), **16** (1964) 209–225.

127. K. Ikeno. Correction: 'Summability methods of Borel type and Tauberian series'. *Tôhoku Math. J.* (2), **19** (1967) 101.

128. A.E. Ingham. On the high-indices theorem for Borel summability. In *Number Theory and Analysis (Papers in Honor of Edmund Landau)*, pages 119–135, Plenum, New York, 1969.

129. K. Ishiguro. On the summability method (Y). *Proc. Japan Acad.* **40** (1964) 482–486.

130. K. Ishiguro. A Tauberian theorem for (J, p_n) summability. *Proc. Japan Acad.* **40** (1964) 870–812.

131. R.B. Israpilov. A certain generalization of Borel's exponential method of summation. *Vestnik Moskov. Univ. Ser. I Mat. Meh.*, **24** (1969) 9–14. Russian. English summary.

132. F.H. Jackson. Borel's integral and q-series. *Proc. Royal Soc. Edinburgh*, **30** (1910) 378–385.

133. F.H. Jackson. The q-integral analogous to Borel's integral. *Messenger*, **47** (1917) 57–64.

134. S.P. Jain. Taylor's series and Borel's polygon of summability. *Proc. Acad. Sci., Allahabad*, **1** (1936) 1–6.

135. R. Jajte. On the compositions of integral means with Borel methods of summability. *Ann. Polon. Math.*, **14** (1963-64) 101–116.

136. R. Jajte. General consistency theorem for Nörlund-Borel methods of summability. *Bull. Acad. Polon. Sci. Sér. Sci. Math. Astronom. Phys.*, **23** (1975) 415–416. Russian summary.

137. A. Jakimovski (Amir). Some relations between the methods of summability of Abel, Borel, Cesàro, Hölder and Hausdorff. *J. Analyse Math.*, **3** (1954) 346–381.

138. A. Jakimovski (Amir). Tauberian constants for the $[J, f(x)]$ transformation. *Pacific J. Math.*, **12** (1962) 567–576.

139. R. san Juan. Ein Algorithmus für die Summation divergenter Reihen. *Union Mat. Argentina*, Publ, **21** (1941) 6 pp. Spanish.

140. W.B. Jurkat. Ein Funktionentheoretischer Beweis für O-Taubersätze bei den Verfahren von Borel und Euler–Knopp. *Arch. Math.*, **7** (1956) 278–283.

141. M.L. Kales. Tauberian theorems related to Borel and Abel summability. *Duke Math. J.*, **3** (1937) 647–666.

142. G. Kangro. Verallgemeinerte Theorie der absoluten Summabilität der divergenten Potenreihen. *Acta Comment. Univ. Tartu* A, **37** (1942) 100 pp.

143. T. Kano. Remark to a Tauberian theorem for Borel summability. *J. Fac. Sci. Shinshu Univ.*, **7** (1972) 5–6.

144. J. Karamata. Application de quelques théorèmes d'inversion à la sommabilité exponentielle. *C. R. Acad. Sci. Paris Sér. I Math.*, **193** (1931) 1156–1159.

145. J. Karamata. Compléments au théorème d'Abel et son inverse. *Rev. math. Un. interbalkanique*, **1** (1937) 161–165.

146. J. Karamata. Un théorème sur le procédé de sommabilité de Borel. *Publ. Math. Univ. Belgrade*, **6/7** (1938) 204–208.

147. J. Karamata. Über die Indexverschiebung beim Borelschen Limitierungsverfahren. *Math. Z.*, **45** (1939) 635–641.

148. N. Khatoon and A. Wafi. On the approximation of function by Borel matrix. *Mat. Vesnik*, **6(19)(34)** (1982) 195–199.

149. E.M. Kil′berg and V.S. Postina. Application of Borel's integral method of summation to a series connected with a Fourier series. In *Studies in current problems of summation and approximation of functions and their applications*, pages 88–94, Dnepropetrovsk. Gos. Univ., Dnepropetrovsk, 1983. Russian.

150. J.P. King. Some results for Borel transforms. *Proc. Amer. Math. Soc.*, **19** (1968) 991–997.

151. K. Knopp. Über das Eulersche Summierungsverfahren. *Math. Z.*, **15** (1922) 226–253.

152. K. Knopp. Über das Eulersche Summierungsverfahren, II. *Math. Z.*, **18** (1923) 125–156.

153. K. Knopp. Bemerkung zum Borelschen Limitierungsverfahren. *Rend. Circ. Mat. Palermo* (2), **54** (1930) 331–334.

154. K. Knopp. Einige Bemerkungen zur $A-$ E_k- und B_k-Summierung. *Rend. Circ. Mat. Palermo* (2), **1** (1952) 129–138.

155. T. Kojima. On generalized Toeplitz's theorems on limit and their applications. *Tôhoku Math. J.*, **12** (1917) 291–326.

156. D.D. Kosambi. Classical Tauberian theorems. *J. Indian Soc. Agric. Stat.*, **10** (1958) 141–149.

157. T. Kövari. On the Borel exceptional values of lacunary integral functions. *J. Analyse Math.*, **9** (1961) 71–109.

158. W. Kratz and U. Stadtmüller. Tauberian theorems for Borel-type methods of summability. *Arch. Math.* (Basel), **55** (1990) 465–474.

159. V.K. Krishnan. Gap Tauberian theorem for generalized Borel summability. *Indian J. Math.*, **24** (1982) 99–112.

160. B. Kuttner and M.R. Parameswaran. A Tauberian theorem for Borel summability. *Math. Proc. Cambridge Philos. Soc.*, **102** (1987) 135–138.

161. B. Kwee. An improvement of a theorem of Hardy and Littlewood. *J. London Math. Soc.* (2), **28** (1983) 93–102.

162. B. Kwee. The relation between the Borel and Riesz methods of summability. *Bull. London Math. Soc.*, **21** (1989) 387–393.

163. L. Leau. Sur les points singuliers situés sur le cercle de convergence et sur la sommation des séries divergentes. *C. R. Acad. Sci. Paris Sér. I Math.*, **127** (1898) 607–609.

164. K.C. Lee. Tauberian theorems for Borel-type methods of summability. *Tamkang J. Math.*, **6** (1975) 271–275.

165. L. Lorch. Translativity for strong Borel summability. *Canad. Math. Bull.*, **9** (1966) 639–645.

166. L. Lorch. The Lebesgue constants for Borel summation of Laplace series. *Monatsh. Math.*, **91** (1981) 291–298.

167. L. Lorch and J. Tzimbalario. A quasi-equivalence between Borel summability and convergence for Fourier–Laguerre series at the endpoint. *Proc. Edinburgh Math. Soc.* (2), **22** (1979) 217–226.

168. R.D. Lord. On some relations between the Abel, Borel, and Cesàro methods of summation. *Proc. London Math. Soc.* (2), **38** (1934) 241–256.

169. G.G. Lorentz. Tauberian theorems and Tauberian conditions. *Trans. Amer. Math. Soc.*, **63** (1948) 226–234.

170. G.G. Lorentz. Direct theorems on methods of summability. *Canad. J. Math.*, **1** (1949) 305–319.

171. G.G. Lorentz. Borel and Banach properties of methods of summation. *Duke Math. J.*, **22** (1955) 129–141.

172. V. Lototskiĭ [V.A. Lotockiĭ]. Conditions for the coincidence of the kernels of the Borel and Euler–Knopp means. *Dokl. Akad. Nauk Ukrain. SSR Ser.A*, 1980, **5** (1980) 16–19,92. Russian. English summary.

173. V. Lototskiĭ [V.A. Lotockiĭ]. Cores of Borel means for unbounded sequences. *Tartu Riikl. Ül. Toimetised*, **504** (1981) 99–104. Russian. English summary.

174. V. Lototskiĭ [V.A. Lotockiĭ]. The kernel of Borel means for a class of unbounded sequences. *Tartu Riikl. Ül. Toimetised*, **661** (1984) 76–80. Russian. English summary.

175. W. Luh. Überkonvergenz bei Euler- und Borel-Verfahren. *Mitt. Math. Sem. Gießen Heft*, **92** (1971) 37–50.

176. E. Maillet. Sur les séries divergentes at les équations différentielles. *Ann. de l'Éc. Norm.*, **20** (1903) 487–518.

177. B. Martić. The application of Karamata–Stirling methods to analytical continuation and relation with Borel's method of summation. *Glasnik Mat.-Fiz. Astronom. Društvo Mat. Fiz. Hrvatske Ser.II*, **16** (1961) 171–175. Serbo-Croatian summary.

178. A.A. Melencov and È.B. Muraev. On the theory of summation of double series by Borel's methods. *Dokl. Akad. Nauk SSSR*, **130** (1960) 1193–1195. Russian. Translated as *Soviet Math. Dokl.*, volume 1, pages 150-152.

179. V.I. Mel′nik. A 'high index' Tauberian theorem for the method of Borel. In *First Republ. Math. Conf. of Young Researchers, Part II*, pages 466–476, Akad. Nauk Ukrain. SSR Inst. Mat., Kiev, 1965. Russian.

180. V.I. Mel′nik. The (B)-property of Borel methods for the summation of series, and theorems of Tauberian type. *Ukrain. Mat. Ž.*, **17** (1965) 64–76. Russian.

181. V.I. Mel′nik. The Tauberian theorem of 'large exponents' for the method of Borel. *Mat. Sb. (N.S.)*, **68** (110) (1965) 17–25. Russian.

182. V.I. Mel′nik. A Tauberian theorem for the Borel summability method. *Izv. Vysš. Učebn. Zaved. Matematika*, 1971 **(114)** (1971) 85–92.

183. V.I. Mel′nik. Tauberian theorems for summability methods of the type of Borel's method. *Ukrain. Mat. Ž.*, **30** (1978) 176–184, 282. Russian.

184. W. Meyer-König. Die Umkehrung des Euler-Knoppschen und des Borelschen Limitierungsverfahrens auf Grund einer Lückenbedingung. *Math. Z.*, **49** (1943) 151–160.

185. W. Meyer-König. Untersuchungen über einige verwandte Limitierungsverfahren. *Math. Z.*, **52** (1949) 259–304.

186. W. Meyer-König. Bemerkung zu einem Lückenumkehrsatz von H.R.-Pitt. *Math. Z.*, **57** (1953) 351–352.

187. W. Meyer-König and K. Zeller. Lückenumkehrsätze und Lückenperfektheit. *Math. Z.*, **66** (1956) 203–224.

188. W. Meyer-König and K. Zeller. On Borel's method of summability. *Proc. Amer. Math. Soc.*, **11** (1960) 307–314.

189. W. Meyer-König and K. Zeller. FK-Räume und Lückenperfektheit. *Math. Z.*, **78** (1962) 143–148.

190. W. Meyer-König and K. Zeller. Euler–Knopp- und Borel–Verfahren komplexer Ordnung. *Math. Z.*, **82** (1963) 394–402.

191. G. Mignosi. Sulle medie di Doetsch delle funzioni. *Note e memorie di mat.*, **1** (1921) 136–158.

192. B.P. Mishra. Strong summability of infinite series on a scale of Abel type summability methods. *Math. Proc. Cambridge Philos. Soc.*, **63** (1967) 119–127.

193. B.P. Mishra. Absolute summability of infinite series on a scale of Abel type summability methods. *Math. Proc. Cambridge Philos. Soc.*, **64** (1968) 377–387.

194. G. Mittag-Leffler. Sur la reprèsentation analytique d'une branche uniforme d'une fonction monogène. *Acta Math.*, **29** (1904) 101–181.

195. R. Mohanty. On the summation of Fourier series by Borel's absolute integral method. *J. London Math. Soc.* (1), **38** (1963) 375–380.

196. D.S. Morse. Relative inclusiveness of certain definitions of summability. *Amer. J. Math.*, **45** (1923) 259–285.

197. R. Mohanty. The absolute summability by Borel's integral method of the allied series of a Fourier series. *J. London Math. Soc.* (1), **38** (1963) 381–384.

198. È.B. Muraev. A generalization of Borel summation methods. *Izv. Vysš. Učebn. Zaved. Matematika*, 1962 **(27)** (1962) 101–108. Russian.

199. È.B. Muraev. Borel summability of a power series on the boundary of the circle of convergence. *Izv. Vysš. Učebn. Zaved. Matematika*, 1962 **(26)** (1962) 110–113. Russian.

200. P. Nalli. Sulle serie di Dirichlet. *Rend. Circ. Mat. Palermo*, **40** (1915) 44–70, 167–168.

201. N. Obrechkoff [N. Obreschkoff]. Sur la sommation de la série de Taylor sur le contour du polygone de sommabilité par la méthode de M. Borel. *C. R. Acad. Sci. Paris Sér. I Math.*, **186** (1928) 1813–1815.

202. N. Obrechkoff [N. Obreschkoff]. Sur la sommation exponentielle de M. Borel. *C. R. Acad. Sci. Paris Sér. I Math.*, **191** (1930) 825–827.

203. N. Obrechkoff [N. Obreschkoff]. Sur quelques généralisations de la sommation de M. Borel. *Annuaire Univ. Sofia, Fac. phys.-math*, **28** (1931) 197–286.

204. N. Obrechkoff [N. Obreschkoff]. Sur une généralisation de la sommation de M. Borel des séries divergentes. *Atti Accad. naz. Lincei, Rend., VI.s.*, **15** (1932) 39–43.

205. N. Obrechkoff [N. Obreschkoff]. Über einige Verallgemeinerungen der Borelschen Summierung der divergenten Reihen. *J. Reine Angew. Math.*, **166** (1932) 208–219.

206. N. Obrechkoff [N. Obreschkoff]. Una generalizzazione della sommazione di Borel. *Rend. Circ. Mat. Palermo* (2), **56** (1932) 449–471.

207. I.I. Ogieveckiĭ [I.I. Ogievec′kiĭ]. On the summability of series by Borel's method of fractional order. *Dopovidi Akad. Nauk Ukraïn. RSR*, **1959** (1959) 815–818. Ukrainian (Russian and English summaries).

208. I.I. Ogieveckiĭ [I.I. Ogievec′kiĭ]. The theory of summing series by Borel's method. *Izv. Vysš. Učebn. Zaved. Matematika*, 1960 **(19)** (1960) 174–183. Russian.

209. I.I. Ogieveckiĭ [I.I. Ogievec′kiĭ]. On the theory of summability of series by Borel's method of fractional order. II. *Dopovidi Akad. Nauk Ukraïn. RSR*, **1962** (1962) 719–722. Ukrainian (Russian and English summaries).

210. I.I. Ogieveckiĭ [I.I. Ogievec′kiĭ]. On the theory of Borel summability of series. II. *Izv. Vysš. Učebn. Zaved. Matematika*, 1964 **(40)** (1964) 100–110. Russian.

211. I.I. Ogieveckiĭ [I.I. Ogievec'kiĭ]. Tauberian theorems for certain methods of summation. *Studies Contemporary Problems Constructive Theory of Functions (Proc. All-Union Conf. Baku)*, **1962** (1965) 377–382. Russian.

212. M.R. Parameswaran. On a comparison between the Cesàro and Borel methods of summability. *J. Indian Math. Soc. (N.S.)*, **22** (1958) 85–92.

213. M.R. Parameswaran. Some remarks on Borel summability. *Quart. J. Math. Oxford Ser.* (2), **10** (1959) 224–229.

214. O. Perron. Zur Theorie der divergenten Reihen. *Math. Z.*, **6** (1920) 158–160.

215. O. Perron. Beitrag zur Theorie der divergenten Reihen. *Math. Z.*, **6** (1920) 286–310.

216. E. Phragmén. Sur le domaine de convergence de l'intégrale infinie $\int_0^\infty F(ax)e^{-a}\,da$, *C. R. Acad. Sci. Paris Sér. I Math.*, **132** (1901) 1396–1399.

217. S. Pincherle. Sugli sviluppi assintotici et le serie sommabili. *Rom. Acc. L. Rend.*, **13** (1904) 513–519.

218. H.R. Pitt. General Tauberian theorems. *Proc. London Math. Soc.*, **44** (1938) 243–288.

219. H.R. Pitt. An elementary proof of the closure in L of translations of e^{-x^2}, and the Borel Tauberian theorem. *Proc. Amer. Math. Soc.*, **8** (1957) 706–707.

220. K. Prasad. On the Borel summability of Legendre series. *Indian J. Pure Appl. Math.*, **4** (1973) 601–605.

221. C.T. Rajagopal. A generalization of Tauber's theorem and some Tauberian constants. *Math. Z.*, **57** (1953) 405–414.

222. C.T. Rajagopal. On a theorem connecting Borel and Cesàro summabilities. *J. Indian Math. Soc. (N.S.)*, **24** (1961) 433–442.

223. C.T. Rajagopal. Gap Tauberian theorems on oscillation for the Borel method (B). *J. London Math. Soc.* (1), **44** (1969) 41–51(1968).

224. C.T. Rajagopal and T. Vijayaraghavan. One-sided Tauberian theorems for Borel, Abel and Riemann-second-order transforms. *Rend. Circ. Mat. Palermo* (2), **4** (1955) 309–322.

225. C.T. Rajagopal and T. Vijayaraghavan. On two Tauberian theorems for the Borel transform of a sequence. *Proc. Indian Acad. Sci. Sect. A.*, **43** (1956) 163–172.

226. M.S. Rangachari. On gap Tauberian theorems for Abel, Borel and Lambert summabilities. *Publ. Ramanujan Inst. No.1*, **1968/69** (1968/69) 269–281.

227. J. Rey Pastor. Zur Theorie der divergenten Reihen. *Tôhoku Math. J.*, **36** (1932) 73–77.

228. J. Rey Pastor. Sur l'application de la méthode de Borel aux séries qui ont des termes nuls. *C. R. Acad. Sci. Paris Sér. I Math.*, **197** (1933) 973–974.

229. E. Ricotti. Sulle serie divergenti sommabili. *Batt. G.*, **48** (1910) 79–111.

230. L.A. Rubel. Summability theory: a neglected tool of analysis. *Amer. Math. Monthly*, **96** (1989) 421–423.

231. C. Ryll-Nardzewski. The Borel method of limitation is perfect. *Bull. Acad. Polon. Sci. Sér. Sci. Math. Astronom. Phys.*, **10** (1962) 649–650.

232. B.N. Sahney. On the Borel summability of the derived Fourier series. *Bull. Calcutta Math. Soc.*, **54** (1962) 211–218.

233. G. Sannia. Sul metodo di Borel per la sommazione delle serie. *Rom. Acc. L. Rend.* (5), **26** (1917) 162–167.

234. G. Sannia. Generalizzazione del metodo di Borel per la sommazione delle serie. *Rom. Acc. L. Rend.* (5), **26** (1917) 603–606.

235. G. Sannia. Le serie di funzioni sommate col metodo di Borel generalizzato. *Rom. Acc. L. Rend.* (5), **26** (1917) 77–81.

236. G. Sannia. Nuovo metodo di sommazione delle serie: estensione del metodo di Borel. *Rend. Circ. Mat. Palermo* (2), **42** (1917) 603–606.

237. G. Sannia. Le serie di potenze di una variabile sommate col metodo di Borel generalizzato. *Torino Atti*, **53** (1917–8) 135–148, 192–206. Nota I, II.

238. G. Sannia. Sulle serie di potenze sommate col metodo di Borel generalizzato. *Rom. Acc. L. Rend.* (5), **27** (1918) 139–142. Nota II.

239. G. Sannia. Estensione e studio di un metodo di sommazione generico di Borel. *Scritti matematici offerti ad Enrico d'Ovidio*, **1918** (1918) 227–252.

240. G. Sannia. Sulle serie di potenze di una variabile sommate col metodo di Borel generalizzato. *Rom. Acc. L. Rend.* (5), **27** (1918) 98–102; 24–27. Nota I, III.

241. G. Sannia. Serie assolutamente sommabili col metodo di Borel generalizzato. *Torino Atti*, **56** (1922) 34–40.

242. R. Schmidt. Die Umkehrsätze des Borelschen Summierungsverfahren. *Schriften Königsberg*, **1** (1925) 205–256.

243. I. Schur. Über lineare Transformationen in der Theorie der unendlichen Reihen. *J. Reine Angew. Math.*, **151** (1920) 79–111.

244. G. Scorza Dragoni. Intorno alla moltiplicazione delle serie che convergono condizionatamente. I, II. *Rom. Acc. L. Rend.* (6), **18** (1933) 193–197, 271–275.

245. M. Servant Essai sur les séries divergentes. *Toulouse Ann.*, **1** (1899) 117–175.

246. B.L.R. Shawyer. On the relation between the Abel and Borel-type methods of summability. *Proc. Amer. Math. Soc.*, **22** (1969) 15–19.

247. B.L.R. Shawyer. Some relations between strong and ordinary Borel-type methods of summability. *Math. Z.*, **109** (1969) 115–120.

248. B.L.R. Shawyer. General absolute Borel-type methods. *J. Reine Angew. Math.*, **245** (1970) 81–90.

249. B.L.R. Shawyer. Abelian theorems for Borel-type methods of summability. *J. London Math. Soc.* (2), **3** (1971) 169–177.

250. B.L.R. Shawyer. On the product of the Cesàro and Borel-type methods of summability. *J. London Math. Soc.* (2), **7** (1973) 111–116.

251. B.L.R. Shawyer and G.S. Yang. On the relation between the Abel-type and Borel-type methods of summability. *Proc. Amer. Math. Soc.*, **26** (1970) 323–328.

252. B.L.R. Shawyer and G.S. Yang. Tauberian relations between the Abel-type and the Borel-type methods of summability. *Manuscripta Math.*, **5** (1971) 341–357.

253. S. Sherif. A Tauberian relation between the Borel and the Lototsky transforms of series. *Pacific J. Math.*, **27** (1968) 145–154.

254. A. Sidi. Borel summability and converging factors for some everywhere divergent series. *SIAM J. Math. Anal.*, **17** (1986) 1222–1231.

255. A.N. Singh and S.P. Khari Borel summability of Jacobi series. *Proc. Nat. Acad. Sci. India* (A), **57** (1987) 743–746.

256. V. Singh and S.R. Sinha. On the Borel summability of the double Fourier series. *Indian J. Math.*, **15** (1973) 197–216.

257. Y. Sitaraman. On Tauberian theorems for the S_α-method of summability. *Math. Z.*, **95** (1967) 34–49.

258. K.M. Slīpenčuk [K.M. Slepenčuk]. A Tauberian theorem for absolute summability by the Borel method. *Dopovidi Akad. Nauk Ukraïn. RSR*, **1966 (186)** (1966) 722–725. Ukrainian. Russian and English summaries.

259. K.M. Slīpenčuk [K.M. Slepenčuk]. A certain general Tauberian type theorem for absolute summability of series, and its application to the Borel method. *Izv. Vysš. Učebn. Zaved. Matematika*, 1969 **(82)** (1969) 61–65. Russian.

260. K.M. Slīpenčuk [K.M. Slepenčuk]. Theorems of Tauberian type for the case of summation of double series by the Borel method. *Ukrain. Mat. Zh.*, **39** (1987) 799–801, 817. Russian.

261. K.M. Slīpenčuk [K.M. Slepenčuk]. Theorems of Tauberian type for matrix transformations and their applications to Borel's method. *Izv. Vysš. Učebn. Zaved. Matematika*, **1977 (186)** (1977) 100–105. Russian.

262. A.D. Sokal. An improvement of Watson's theorem on Borel summability. *J. Math. Phys.*, **21** (1980) 261–263 (this is also cited in the Applied Bibliography).

263. A.S. Sokolin. Über zwei Klassen von Summationsmethoden für divergente Reihen. *Uspechi mat Nauk.*, **12** (1957) 381–384. Russian.

264. R.P. Soni and K. Soni. On a class of Borel-type summability methods. *J. Math. Phys. Sci.*, **21** (1987) 567–583.

265. M. Stieglitz. Die allgemaine Form der *o*-Tauber-Bedingung für das Euler-Knopp- und Borel-Verfahren. *J. Reine Angew. Math.*, **246** (1971) 172–179.

266. F. Strasser. Über die Verträglichkeit der Verfahren $E_p B_q$ und S_α der Limitierungstheorie. In *Von der Technischen Hochschule Stuttgart zur Erlangung der Würde eines Doktors der Naturwissenschaften*

genehmigte Abhandlung, Technische Hochschule Stuttgart, Stuttgart, 1967.

267. S.P.M. Strube. Tauberian remainder theorems for the Borel methods. *J. Reine Angew. Math.*, **273** (1975) 125–133.

268. J. Swetits. Generalized Borel summability of Fourier series. *J. Math. Anal. Appl.*, **30** (1970) 308–312.

269. O. Szasz. On products of summability methods. *Proc. Amer. Math. Soc.*, **3** (1952) 257–263.

270. O. Szasz. On the product of two summability methods. *Ann. Soc. Polon. Math.*, **25** (1953) 75–84.

271. N. Takagi. On Borel's method of summation. *Sci. Rep. Tôhoku Univ., I.s.*, **24** (1935) 45–54.

272. L. Tam. Cesàro and general Euler–Borel summability. *Proc. Amer. Math. Soc.*, **119** (1992) 747–755.

273. L. Tam. A Tauberian theorem for the general Euler–Borel summability method. *Canad. J. Math.*, **44** (1992) 1100–1120 (English summary).

274. J.B. Tatchell. A theorem on Riesz summability. *J. London Math. Soc.*, **29** (1954) 49–59.

275. J. Teghem. Sur les conditions d'applicabilité d'une méthode de prolongement analytique de Borel. *Acad. Roy. Belgique. Bull. Cl. Sci.* (5), **35** (1949) 177–185.

276. J. Teghem. Addenda à la note: 'Sur les conditions d'applicabilité d'une méthode de prolongement analytique de Borel'. *Acad. Roy. Belgique. Bull. Cl. Sci.* (5), **37** (1951) 20.

277. J. Teghem. Sur des extensions d'une méthode de prolongement analytique de Borel. In *Colloque sur la théorie des suites, tenu à Bruxelles du 18 au 20 Décembre, 1957*, pages 87–95, Centre Belge de Recherches Mathématiques, Libraire Gauthier-Villars, Paris; Établissements Ceuterick, Louvain, 1958.

278. G. Tenenbaum. Sur le procédé de sommation de Borel et la répartition du nombre des facteurs premiers des entiers. *L'Enseignement Mathématique, Series 2*, **26** (1980) 225–245.

279. V. Thiruvenkatacharya. On Borel's method for double series. *J. Indian Math. Club*, **16** (1926) 155–163.

280. O. Toeplitz. Über allgemeine lineare Mittelbildungen. *Prace mat. -fiz.*, **22** (1911) 113–119.

281. N.V. Tret′yak. A property of the Borel matrix. *Teor. Funktsiĭ Funktsional. Anal. i Prilozhen.*, **53** (1990) 51–56. Russian.

282. L.M. Tripathi and M. Tripathi. On strong Borel summability of Legendre series. *Proc. Math. Soc.*, **6** (1990) 53–58.

283. L.M. Tripathi and V.P. Yadav. On Borel summability of Legendre series. *Proc. Nat. Acad. Sci. India Sect. A.*, **57** (1987) 320–323.

284. G. Valiron. Remarques sur la sommation des séries divergentes par les méthodes de M. Borel. *Rend. Circ. Mat. Palermo* (2), **42** (1917) 267–284.

285. G. Valiron. Sur les directions de Borel de certaines fonctions entières d'ordre infini. *C. R. Acad. Sci. Paris Sér. I Math.*, **194** (1932) 1552–1555.

286. G. Valiron. Sur les directions de Borel de certaines fonctions entières. *C. R. Acad. Sci. Paris Sér. I Math.*, **194** (1932) 1305–1308.

287. G. Valiron. Méthodes de sommation et directions de Borel. *Ann. Scuola norm. super. Pisa, II.s.*, **2** (1933) 355–380.

288. J.C. Vignaux. Über absolut summierbare Reihen nach einer verallgemeinerten Borelschen Methode. *Contrib. Estud. Ci. fís. y mat.*, **5** (1931) 277–289.

289. J.C. Vignaux. Über die Verallgemeinerung eines Borelshen Lemmas. *Contrib. Estud. Ci. fís. y mat.*, **5** (1931) 323–329.

290. J.C. Vignaux. Sobre series sumables absolutamente con el método generalizando de Borel. *Publicaciones La Plata*, **93** (1931) 279–289.

291. J.C. Vignaux. Einige Sätze über das Produkt B'-summierbarer Reihen. *An. Soc. Ci. Argent.*, **115** (1933) 25–27. Spanish.

292. J.C. Vignaux. Reihen, die mit der verallgemeinerten Le-Royschen Methode summiert werden können. *An. Soc. Ci. Argent.*, **116** (1933) 42–43. Spanish.

293. J.C. Vignaux. Sur le produit de séries sommables par la méthode de Borel. *Bull. Sci. math., II.s.*, **57** (1933) 211–219.

294. J.C. Vignaux. Sur la sommabilité totale par la méthode de M. Borel. *Atti Accad. naz. Lincei, Rend., VI.s.*, **17** (1933) 538–540.

295. J.C. Vignaux. Sur une généralisation de la sommation de M. Borel. *C. R. Acad. Sci. Paris Sér. I Math.*, **197** (1933) 668–670.

296. J.C. Vignaux. Über die oszillierende Reihe von Borel. *An. Acad. Nac. Ci. exact. Fis. y Nat., Buenos Aires*, **3** (1933) 155–156. Spanish.

297. J.C. Vignaux. Über B'-oszillierende Reihen. *An. Soc. Ci. Argent.*, **116** (1933) 41–42. Spanish.

298. J.C. Vignaux. Quelques théorèmes sur le produit des séries sommables par la méthode de M. Borel. *Tôhoku Math. J.* (1), **40** (1935) 79–84.

299. J.C. Vignaux. Erweiterung der Borelschen Summationsmethode auf Reihen von Funktionen einer komplexen dualen hyperbolischen Variablen. *An. Soc. Ci. Argent.*, **122** (1936) 193–231. Spanish.

300. J.C. Vignaux. Sätze über die verallgemeinerte Borelsche Summationsmethode. *Contrib. Estud. Ci. fís. mat.*, **1** (1938) 543–544. Spanish.

301. T. Vijayaraghavan. A theorem concerning the summability of series by Borel's method. *Proc. London Math. Soc.* (2), **27** (1928) 316–326.

302. T. Vijayaraghavan and C.T. Rajagopal. On two Tauberian theorems for the Borel transform of a sequence. *Proc. Indian Acad. Sci. Sect. A.*, **43** (1956) 163–172.

303. E.B. van Vleck. Selected topics in the theory of divergent series and continued fractions. Boston Coll. Lect., (1903) 75–77, 95–107, 123–134.

304. A. Wafi and N. Khatoon. On the approximation of function by Borel matrix. *Mat. Vesnik*, **6(19)(34)** (1982) 195–199.

305. G.N. Watson A theory of asymptotic series. *Philos. Trans. Roy. Soc. London* A, **211** (1912) 279–313.

306. N. Weiner. Tauberian theorems. *Annals of Math.* (2), **33** (1932) 1–100.

307. N. Weiner and W.T. Martin. Taylor's series of entire functions of smooth growth. *Duke Math. J.*, **3** (1937) 213–223.

308. J.M. Whittaker. The absolute summability of Fourier series. *Proc. Edinburgh Math. Soc.* (2), **2** (1930–31) 1–5.

309. J.M. Wills. Note zu einem Borelschen Summationsverfahren. *Math. Z.*, **92** (1966) 323–330.

310. L. Włodarski. Propriétés des méthodes continues de limitation du type de Borel. *Bull. Acad. Polon. Sci. Cl. III.*, **4** (1956) 173–175.

311. L. Włodarski. Sur la concordance entre les méthodes intégrales de sommation du type de Borel. *Bull. Acad. Polon. Sci. Cl. III.*, **4** (1956) 177–178.

312. L. Włodarski. Sur les méthodes continues de limitation du type de Borel. *Ann. Polon. Math.*, **4** (1958) 137–164.

313. L. Włodarski. General method of limitation of the Borel type. *Bull. Acad. Polon. Sci., Sér. Sci. math. astron. phys.*, **7** (1959) 199–200.

314. L. Włodarski. On some properties of Borelian methods of the exponential type. *Ann. Polon. Math.*, **10** (1961) 177–196.

315. S.P. Yadav. On Borel-summability of Jacobi series. *Vijnana Parishad Anusandhan Patrika*, **19** (1976) 71–76. Hindi. English summary.

316. G.S. Yang. A sufficient condition for Borel-type methods of summability. *Soochow J. Math. Natur. Sci.*, **1** (1975) 67–71. (Chinese summary) Collection of articles in commemoration of the 75th anniversary of the founding of Soochow University.

317. K. Zeller. Sur le méthode de sommation d'Abel. *C. R. Acad. Sci. Paris Sér. I Math.*, **236** (1953) 568–569.

318. A. Zygmund. Über einige Sätze aus der Theorie der divergenten Reihen. *Bull. Acad. Polon.* (A), (1927) 309–331.

319. A. Zygmund. On a theorem of Ostrowski. *J. London Math. Soc.*, **6** (1931) 162–163.

Bibliography

APPLIED ARTICLES

1. S. Albeverio; B. Tirozzi and B. Zegarlinski. Rigorous results for the free-energy in the Hopfield model. *Communications in Mathematical Physics*, **150** (1992) 337-373.

2. H. Aoyama and A.M. Tamra. Path integral for quantum tunnelling. *Nuclear Physics B* , **384** (1992) 229–250.

3. G. Auberson. Borel summability for a nonpolynomial potential. *Comm. Math. Phys.*, **84** (1982) 531–546.

4. G. Auberson and G. Mennessier. Some properties of Borel summable functions. *J. Math. Phys.*, **22** (1981) 2472–2481.

5. G. Auberson and G. Mennessier. The reciprocal of a Borel summable function is Borel summable. *Comm. Math. Phys.*, **100** (1985) 439–446.

6. M. Bergère and F. David Ambiguities of renormed ϕ_4^4 field theory and the singularities of its Borel transfer. *Phys. Lett.*, **135 B** (1984) 412–416.

7. C. Billionnet and P. Renouard. Analytic interpolation and Borel summability of the $\frac{l}{N}|\phi|_4^4$ models. *Comm. Math. Phys.*, **84** (1982) 251–295.

8. E. Brézin, J.C. LeGuillon, and J. Zinn-Justin. Perturbation theory at large order I. The ϕ^{2N} interaction. *Phys. Rev. D.* **15**, 1977, 1544–1557.

9. E. Brézin, J.C. Guillou, and J. Zinn-Justin. Perturbation theory at large order II. Role of the vacuum instability. *Phys. Rev. D*, **15** , 1977, 1558–1564.

10. E. Brézin, G. Parisi, and J. Zinn-Justin. Perturbation theory at large orders for a potential with degenerate minima. *Phys. Rev. D*, **16** , 1977, 408–412.

11. C. de Calan and V. Rivasseau. Local existence of the Borel transform in Euclidean ϕ_4^4. *Comm. Math. Phys.*, **82** (1981) 69–100.

12. C. de Calan and V. Rivasseau. A comment on the local existence of the Borel transform in Euclidean ϕ_4^4. *Comm. Math. Phys.*, **91** (1983) 265.

13. E. Caliceti, V. Grecchi, and M. Maioli. The distributional Borel summability and the large coupling ϕ^4 lattice fields. *Comm. Math. Phys.*, **104** (1986) 163–174. See erratum.

14. E. Caliceti, V. Grecchi, and M. Maioli. Erratum:'the distributional Borel summability and the large coupling ϕ^4 lattice fields'. *Comm. Math. Phys.*, **113** (1987) 173–176.

15. S. Chadha and P. Olesen. On Borel singularities in quantum field theory. *Phys. Lett.*, **72 B** (1977) 87–90.

16. F. Constantinescu, K. Klöckner, and U. Scharffenberger. Borel summability in the disorder parameter of the averaged Green's function for Gaussian disorder. *Comm. Math. Phys.*, **98** (1985) 203–211.

17. W.Y. Crutchfield II. Method for Borel-summing instanton singularities. Introduction. *Phys. Rev.*, **19 D** (1979) 2370-2384.

18. E. Cunningham. An extension of Borel's exponential method of summation of divergent series applied to linear differential equations. *Proc. London Math. Soc.* (1), **3** (1905) 157–169.

19. A. Duncan and H.F. Jones. Convergence proof for optimized delta-expansion anharmonic-oscillator. *Physical Review*, **47** (1993) 2560–2572.

20. J.-P. Eckmann and H. Epstein. Time-ordered products and Schwinger functions. *Comm. Math. Phys.*, **64** (1979) 95–130.

21. J.-P. Eckmann and H. Epstein. Borel summability of the mass and the s-matrix in ϕ^4 models. *Comm. Math. Phys.*, **68** (1979) 245–258.

22. J.-P. Eckmann, J. Magnen, and R. Sénéor. Decay properties and Borel summability for the Schwinger functions in $P(\phi)_2$ theories. *Comm. Math. Phys.*, **39** (1975) 215–271.

23. J. Frölich, A. Mardin, and V. Rivasseau. Borel summability of the $1/N$ expansion for the N-vector 2 $[O(N)$ non-linear $\sigma]$ models. *Comm. Math. Phys.*, **86** (1982) 87–110.

24. K. Gawędzki, A. Kupiainen, and B. Tirozzi. Borel summability of the perturbation series in a hierarchical $\lambda(\nabla\phi)^4$ model. *J. Statist. Phys.*, **36** (1984) 145–162.

25. D.Z. Goodson, M. Lopezcabrera, D.R. Herschbach, and J.D. Morgan. Large-order dimensional perturbation-theory for 2 electron atoms. *Journal of Chemical Physics*, **97** (1992) 8481–8496.

26. S. Graffi and V. Grecchi. Borel summability and indeterminacy of the Stieltjes moment problem: application to the anharmonic oscillators. *J. Math. Phys.*, **19** (1978) 1002–1006.

27. S8. Graffi and V. Grecchi. Borel summability of the unequal double-well. *Comm. Math. Phys.* **92** , 1984, 397–403.

28. S. Graffi and V. Grecchi. The Borel sum of the double-well perturbation series and the Zinn-Justin conjecture. *Phys. Lett.*, **121 B** (1983) 410–414.

29. S. Graffi, V. Grecchi, and B. Simon. Borel summability: application to the anharmonic oscillator. *Phys. Lett.*, **32 B** (1970) 631–634.

30. V. Grecchi and M. Maioli. Borel summability beyond the factorial growth. *Ann. Inst. H. Poincaré Phys. Théor.*, **41** (1984) 37–47. French summary.

31. V. Grecchi and M. Maioli. Generalized logarithmic Borel summability. *J. Math. Phys.*, **25** (1984) 3439–3443.

32. V. Hakim and K.Mallick. Exponentially small splitting of seperatrices, matching in the complex-plane and Borel summation. *Nonlinearity*, **6** (1993) 57–70.

33. F.J. Halfkann and V. Dohm. Minimal renormalization without epsilon-expansion – amplitude functions in 3 dimensions below to *Zeitschrift Fur Physik B-Condensed Matter*, **89** (1992) 79–87.

34. I.W. Herbst. Stark effect revisited. *Phys. Rev. Lett.* **41** , 1978, 67–69.

35. I.W. Herbst and B. Simon. Dilation analyticity in constant electric field. *Comm. Math. Phys.* **80** , 1981, 181–216

36. G. 't Hooft. Borel summability of a four-dimensional Euclidean field theory. *Phys. Lett.*, **119 B** (1982) 369–371.

37. G. 't Hooft. On the convergence of a planar diagram expansion. *Comm. Math. Phys.*, **86** (1982) 449–464.

38. G. 't Hooft. Rigorous construction of planar field theories in four-dimensional Euclidean space. *Comm. Math. Phys.*, **88** (1983) 1–25.

39. N.N. Khuri. Borel summability and the renormalization group. *Phys. Rev.*, **16 D** (1977) 1754–1761.

40. N.N. Khuri. Zeros of the Gell-Mann–Low function and Borel summation in renormalization theories. *Phys. Lett.*, **82 B** (1979) 83–88.

41. C. Kopper. On the Borel summability of planar perturbation series: transition to Minkowski space. *Comm. Math. Phys.*, **116** (1988) 57–72.

42. B. Lautrup. On high order estimates in QED. *Phys. Lett.* **69 B** , 1977, 109–111.

43. J.C. Le Guillou and J. Zinn-Justin. Critical exponents for the n-vector model in three dimensions from field theory. *Phys. Rev. Lett.* **39** , 1977, 95–98.

44. L.N. Lipatov. Divergence of the perturbation-theory series and pseudoparticles. *Zh. Eksp. Teor. Fiz. Pizma.* **25** , 1977, 116-119.

45. L.N. Lipatov. Divergence of the perturbation-theory series and the quasiclassical theory. *Zh. Eksp. Teor. Fiz.* **72**, 1977, 411–427.

46. J. Magnen and R. Sénéor. Phase space cell expansion and Borel summability for the Euclidean ϕ_3^4 theory. *Comm. Math. Phys.*, **56** (1977) 237–276.

47. M. Maioli. Borel-Le Roy summability of the high temperature expansion for classical continuous systems. *Comm. Math. Phys.*, **98** (1985) 245–251.

48. M.F. Marziani. A connection between Borel and Padé summation techniques. *Lett. Nuovo Cimento* (2), **37** (1983) 124–128.

49. M.F. Marziani. Convergence of a class of Borel-Padé-type approximants. *Nuovo Cimento B* (11), **99** (1987) 145–154. Italian and Russian summaries.

50. A. Moroz. Strong asymptotic conditions (short guide to using summability methods). *Czechoslovak Journal of Physics*, **42** (1992) 753–763.

51. G. Parisi. Asymptotic estimates in perturbation theory. *Phys. Lett.* **66 B** , 1977, 167–169.

52. G. Parisi. Singularities of the Borel transform in renormalizable theories. *Phys. Lett.*, **76 B** (1978) 65–66.

53. G. Parisi. The Borel transform and the renormalizable group. *Phys. Rep.*, **49** (1979) 215–219.

54. M. Reeken. Summability methods in perturbation theory. *J. Math. Phys.*, **11** (1970) 822–824.

55. P. Renouard. Analyticité et sommabilité 'de Borel' des fonctions de Schwinger du modèle de Yukawa en dimension $d = 2$. I. *Ann. Inst. H. Poincaré Phys. Théor.*, **27** (1977) 237–277.

56. P. Renouard. Analyticité et sommabilité 'de Borel' des fonctions de Schwinger du modèle de Yukawa en dimension $d = 2$. II. *Ann. Inst. H. Poincaré Phys. Théor.*, **31** (1979) 235–318.

57. V. Rivasseau. Construction and Borel summability of planar 4-dimensional Euclidean field theory. *Comm. Math. Phys.*, **95** (1984) 445–486.

58. V. Rivasseau. Rigorous construction and Borel summability for a planar four dimensional field theory. *Phys. Lett.*, **137 B** (1984) 98–102.

59. V. Rivasseau and E. Speer. The Borel transform in Euclidean ϕ^4_ν; Local existence for $\Re\nu < 4$. *Comm. Math. Phys.*, **72** (1980) 293–302.

60. L. Rosen and B. Simon. The $(\phi^{2n})_2$ field hamiltonian for complex coupling constant. *Trans. Amer. Math. Soc.* **165** , 1972, 365–379.

61. B. Simon. Borel summability of the ground-state energy in spatially cutoff $(\phi^4)_2$. *Phys. Rev. Lett.*, **25** (1970) 1583–1586.

62. B. Simon. Determination of eigenvalues by divergent perturbation series. *Adv. Math.* **7** , 1971, 240–253.

63. B. Simon. Large orders and summability of eigenvalue perturbation theory: A mathematical overview. *Int. J. Quan. Chem.* **21** , 1982, 3–25.

64. B. Simon. Summability methods, the strong asymptotic condition and unitarity in quantum field theory. *Phys. Rev. Lett.*, **28** (1972) 1145–1146.

65. A.D. Sokal. An improvement of Watson's theorem on Borel summability. *J. Math. Phys.*, **21** (1980) 261–263 (this is also cited under 'Articles').

66. A. Voros. The return of the quartic oscillator. The complex WKB method. *Ann. Inst. H. Poincaré Phys. Théor.* **39** , 1983, 211–338.

67. W. Wagner. Borel-summability of the high temperature expansion for classical continuous systems. *Comm. Math. Phys.*, **82** (1981/82) 183–189.

68. E.J. Weniger; J. Cizek, and F. Vinette. The summation of the ordinary and renormalized perturbation-series for the ground-state energy of

the quartic, sextic, and octic anharmonic-oscillators using nonlinear sequence transformations. *Journal of Mathematical Physics*, **34** (1993) 571–609.

69. N.K. Wilkin and M.A. Moore. Analysis of the perturbation-series for the specific-heat of a thin-film superconductor near HC2. *Physical Review B-Condensed Matter*, **47** (1993) 957–966.

70. V.I. Zakharov. QCD Pertubative expansions in large orders. *Nuclear Physics*, **385** (1992) 452–480.

Index

The manufacturer's authorised representative in the EU for product safety is Oxford University Press España S.A. of el Parque Empresarial San Fernando de Henares, Avenida de Castilla, 2 – 28830 Madrid (www.oup.es/en or product.safety@oup.com). OUP España S.A. also acts as importer into Spain of products made by the manufacturer.

www.ingramcontent.com/pod-product-compliance
Lightning Source LLC
LaVergne TN
LVHW011755200926
848484LV00003B/715

9780198535850